THE
PROPAGATOR'S
HANDBOOK

*Fifty foolproof recipes – hundreds of plants
for your garden*

PETER THOMPSON

Illustrations by Josie Owen

David & Charles

YEAR PLANNER FOR THE RECIPES

The recipes in the following table have been arranged according to the month when they are started. Elsewhere in this book references are made to seasons, rather than to months – this has been done to make the information accessible to gardeners living anywhere in temperate parts of the world.

However, the seasons are less clearly defined and references to them convey less precise information than recommendations to do particular things in particular months. The author lives in the Welsh Marches of the United Kingdom and the timings referred to in the table below are appropriate for that part of the world – lesser or greater adjustments would have to be made elsewhere to take account of different conditions, opportunities and problems.

The relationship between months and seasons in this part of the world are shown below:

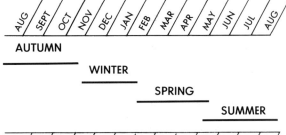

Broad Correspondence for Southern Hemisphere

PLANT (RECIPE NO)	DIVISION	SEEDS	CUTTINGS
Blue Poppies (16/p63)		Jan/Feb	
Primula denticulata (42/p127)			Jan/Mar
Freesias (50/p142)		Feb	
Chrysanthemums (48/p138)			Feb/Mar
Busy Lizzies (43/p130)		Feb/Mar	
Nemesias (13/p56)		Mar/Apr	
Petunia (14/p59)		Apr/May	
Delphinium (3/p33)	Apr/May		
Phlox (border) (26/p87)			Apr/May
Verbena 'Sissinghurst' (46/p135)			Apr/May
Larkspurs (10/p52)		Apr/May	
Lavenders 'Munstead' and 'Old English' (20/p70 & 24/p82)		Apr/May	May
Bamboos (7/p42)	May		
Clematis orientalis (25/p85)			May
Streptocarpus (49/p141)			May/Jun
Columbines (15/p61)		May/Jun	
Wallflowers (11/p54)		May/Jun	
Polyanthus (17/p65)		May/Jun	
Christmas Roses (18/p66)		Jun/Jul	
Bowles's Golden Grass (8/p43)		Jun/Jul	
Spiraea 'Gold Flame' (29/p97)			Jun/Jul
Mock Orange (27/p91)			Jun/Aug
Bluebells (19/p68)		July	
Daffodils (1/p31)	July		

PLANT (RECIPE NO)	DIVISION	SEEDS	CUTTINGS
Japanese azaleas (32/p103)			July
Roses (36/p112)			July
Bergenia 'Sunningdale' (40/p122)			Jul/Aug
Geranium macrorrhizum (30/p99)			Jul/Aug
Hostas (4/p34)	Jul/Aug		
Viola 'Prince Henry' (44/p132)		Jul/Aug	
Daphne mezereum (22/p72)		August	
Erica 'Springwood White' (31/p101)			August
Geraniums (Pelagonium) (45/p133)		Aug/Sept	
Schizanthus (47/p137)		Aug/Sept	
Snapdragons (12/p55)		Aug/Sept	
Penstemon 'Firebird' (34/p108)			Aug/Sept
Ivies (37/p114)			Sept
Oak Trees (23/p75)		October	
Hydrangeas (28/p94)			Oct/Nov
Thuja 'Rheingold' (38/p116)			Oct/Nov
Roses (39/p119)			Oct/Nov
Rhododendron 'Pink Pearl' (6/p39)	Oct/Jan		
Purple Filbert (5/p38)	Oct/Feb		
Rhododendron species/ Hybrids (21/p71)	Oct/Feb		
Ceanothus (33/p106)			Nov
Violas (35/p110)			Nov
Lilies (41/p124)			Nov/Jan
Michaelmas Daisy (2/p32)	Nov/Dec		

A DAVID & CHARLES BOOK

Copyright © Peter Thompson 1993, 1996
Line drawings by Josie Owen

Peter Thompson has asserted his right to be identified as author of this work in accordance with the Copyright, Designs and Patents Act, 1988.

A catalogue record for this book is available from the British Library.

ISBN 0 7153 0426 7

First published 1993
Reprinted 1994, 1995
First paperback edition 1996
Reprinted 1998

Typeset in Garamond Light ITC and Spartan
by ABM Typographics Ltd, Hull
and printed in France by Imprimerie Pollina S.A.
for David & Charles
Brunel House   Newton Abbot   Devon

CONTENTS

1

INTRODUCTION: WHY BOTHER?

Lenten roses are much better garden plants than the well-known Christmas roses. They flower more freely, grow more reliably and are less at risk from pests and diseases. Special forms like 'Trotter's Spotted' will not come true from seed, but all the seedlings will have attractive flowers, and will grow into immensely long-lived plants that will become more impressive as the years pass. Established plants sow themselves in the garden, but new ones can be introduced, or particularly attractive strains increased, by sowing their seeds (recipe 18) and growing on seedlings

Gardening rumours and potting shed whispers may have persuaded you that only experienced Adams and Eves with distinctly green fingers can be successful propagators. But this is a book for beginners as well as established gardeners. It is for people whose gardening has to be fitted into a busy life, and for grey-thumbed Lynns whose efforts seem doomed to end with fallen seedlings and wilted cuttings.

For propagation is not difficult unless we make it so. Plants have done it unaided and successfully for aeons, and we need only provide conditions that let them get on with it. This is a book which will show you how to propagate using simple equipment that works without fuss. Anyone can follow the directions without squandering money, wasting precious time or relying on years of experience.

You will find descriptions of how to propagate plants laid out as clearly and unambiguously as possible, like recipes in a cookery book. The methods described are those which I have used again – and again. Other ways could be used and variations could be introduced, and if you want to try them, do so – it's the best way to learn – but don't ask me to guarantee the results!

Many readers will not be gardeners already equipped with the mental agility and resilience needed to cope with the natural wilfulness of plants. They will be aspiring gardeners. Some will be accountants, local government officials, retired admirals or computer programmers who have learnt to follow rules like guiding tapes, marking a path through minefields of doubt. The ebb and flow of ifs, buts and perhapses that govern everything we do in our gardens may come as a shock to them.

So, each recipe is presented as a single track following a well-marked path to its destination. It tells you the best times of year to do things and the most economical equipment to use. It also tells you how to grow the seedlings and rooted cuttings on afterwards until they are big enough to take their places in the garden.

It is an approach that cannot always work. A thousand interruptions – the weather, the neighbour's cat, one's own children, better things to do, acts of God and attacks by slugs will see to that. But recipes in cookery books do not always work the first time either. The second time – and the third – if they are well conceived, they can be surprisingly successful. The same applies to gardening.

But why bother to produce at home plants that can be bought, in almost bewildering variety, at any garden centre or nursery? It is a question I cannot answer for you. I can only explain my own reasons for doing so. More that 90 per cent of the plants in my garden are home-grown, so there must be compelling reasons somewhere.

None is more compelling than the conclusion reached by simple arithmetic when I estimate the number of plants (something between fifteen and twenty thousand) and multiply by the most notional cost per plant. Any value chosen would convince me that, by growing the plants myself, I have been able to use more, in much more imaginative ways and with far greater freedom, than I could ever have afforded to do if I had bought them.

I have been able to plant mixtures of perennials in large groups to cover the ground beneath trees and among shrubs, and form communities which do much of

my gardening for me. I should never have planted so generously if I had had to load trolleys in garden centres with thirty or forty plants of a kind. A quick tour choosing a plant here, two or three there, to be brought home as stock plants from which to produce more, is not just easier on the pocket – it conforms to the way I like to shop.

I much prefer taking cuttings and sowing seeds to pulling up weeds. Since the plants produced fill spaces that weeds would occupy, propagation is a congenial way to avoid a repetitive and monotonous job. I am no longer bothered about whether to risk growing plants that are not quite hardy. I can take cuttings in the autumn of anything I am doubtful about and shelter them in a frame or greenhouse as a reserve, so that bitter winters do not lead to sad disasters. I can buy a few plants each spring of the colourful things that brighten up a terrace or a yard and multiply them by ten to make an abundant display later in the summer.

My gardening costs less because I grow the plants myself; the material and practical advantages are self-evident. In less definable ways, the garden has become more my own. The plants that grow in it are *my* plants – not the mass productions of anonymous nurseries that previously I used to buy from garden centres.

Magnolia × soulangeana can be propagated from cuttings – but is not easy. A more foolproof method is to make layers by bending shoots down to the ground, and holding them firmly in place while they form roots (recipe 6). Often one or two of the flexible young branches on a newly bought plant can be used as layers without spoiling the shape of it.

5

2

A BRIEF VIEW OF PROPAGATION

Gardening is the craft through which the art of garden-making becomes possible. It combines the skill of choosing plants capable of growing in a particular place with the ability to enable them to do so. This knack of choosing felicitously, combined with good management, is the foundation of successful attempts to propagate plants.

It is far more important than investing in elaborate equipment. Some of the most dependable recipes in this book require nothing more than space in a corner of the garden. Equipment of any kind only works well with skilful management, experience of what can go wrong and time spent making it work. The more complex or precise it is, the greater the demands it makes.

Plants can be propagated in three ways: from seed, by division, and from cuttings.

SEEDS	DIVISION	CUTTINGS
Seeds, naturally produced by plants, can be collected and stored for long periods. New individuals can be produced from them when they are sown in the right conditions.	An extremely simple method in which a plant is divided into a number of separate, more or less self-supporting individuals, each with its own shoots and roots.	Bits of a plant, eg shoots, roots etc, are cut off, and then kept alive under artificial conditions until they regenerate the missing parts necessary for their survival.

Seeds develop after fertilisation of the ovules by pollen, and each seedling has a unique individuality derived from the combination of genes it receives from its parent(s). As a result, plants grown from seeds are genetically different from their parents and from their siblings, even those out of the same seed capsule. Differences between individuals are sometimes barely perceptible, but sometimes very great. The reasons for this are varied and may be complex. Their effect as far as gardeners are concerned, is that seeds cannot always be relied on to produce plants which closely resemble their parent(s).

Nevertheless, seeds are a convenient, regularly used way to propagate plants which do breed true naturally, or which have been carefully bred to do so. Many garden perennials and shrubs have not been much changed from their wild ancestors, and when their seeds are sown there is usually little variation from one seedling to another. Commercial seeds of annual flowers and vegetables provide familiar examples of carefully selected strains in which one plant of a particular kind looks very much like another, and seeds collected from them grow into plants very similar to their parents.

But in other groups of plants, seeds cannot be used so reliably. Many of the more expensive – though not always better – seed strains of flowers and vegetables are produced by a careful blending of two different strains. The offspring, known as F1 hybrids, have a critically composed genetic constitution and display precisely defined and very uniform qualities. But when the flowers of these hybrids are pollinated, the genes are thrown back into the melting pot and the random mixtures which result produce a varied array of good, bad or indifferent plants.

Many of our most popular garden plants – for example, roses and gladioli – have been changed so much in cultivation that they scarcely resemble their wild ancestors. Several different species, hybridised and then selected for many generations, may play a part in their family trees. These highly bred and much-changed plants never breed true from seed, and the odds against a seedling growing up to be as good as, let alone better than, its parent may be hundreds or even thousands to one.

Confronted with such odds, gardeners abandon seeds and sex, and resort to vegetative methods of propagation: they divide plants or take cuttings from them. In other words, they 'clone' them, and so perpetuate unchanged what they already have.

Those venturing to propagate plants for the first time may feel relieved to read that there are only three ways. But there are ways to do the ways, and so complications set in. The recipes in this book provide a guide through this labyrinth but, before setting out to try one rather than another, an overview of the menu with a few comments on the variety of the flavours, might be helpful.

A View of the Recipes

Each recipe describes one way to propagate a particular plant. Sometimes there is more than one recipe for a plant, for example, roses and lavenders. Similar methods could be used to propagate other plants, and a list of their names will be found at the end of the relevant recipe. In the following table the recipes are listed in order according to method: division first, followed by growing from seed, and, finally, cuttings. Some recipes are easy, others more difficult. The column headed Skill Level provides an insight into the level of skill and care needed for success on a scale from 1 to 7 – ranging from 'Super Easy' to 'Testing':

1 Super Easy – almost foolproof.
2 Easy – requires no skill and little attention.
3 Straight-forward – succeeds with a little care and occasional attention.
4 Amenable – good results follow regular care and attention.
5 Responsive – variable results; highly dependent on quality of care.
6 Complex – success depends on achieving a balance between sometimes opposing factors.
7 Testing – only successful with skilled care and frequent attention.

Plant	Recipe	Method	Equipment	Skill Level	Comments
PLANTS FOR THE BORDERS					
Delphinium	3	Division	Outdoors	(2)	A very good way to increase old clumps of delphiniums growing in the garden.
Hostas	4	Division	Cold frame	(3)	A simple method for increasing a very valuable, low-maintenance garden plant.
Michaelmas Daisy	2	Division	Outdoors	(1)	A practical way to increase many herbaceous plants – especially those that are vigorous and easily grown.
Blue Poppies	16	Seed	Greenhouse	(5)	This plant has a reputation for being difficult. Grow it cool, in well-drained compost, and be sure never to let the compost dry out.
Bowles's Golden Grass	8	Seed	Cold frame	(3)	Once in the garden, it will sow itself.
Columbine	15	Seed	Outdoors	(2)	This, and many other perennials, can easily be grown from seed in an outdoor seedbed.
Hellebores ('Christmas Roses')	18	Seed	Cold frame	(6)	Takes several years to produce flowering plants, and the young plants will need skilful care.
Larkspur	10	Seed	Outdoors	(2)	Main problems are weeds and wet weather.
Nemesias	13	Seed	Greenhouse	(4)	Provided the seedlings can be given a small amount of attention daily, these are not hard to grow and are very economical to produce.
Petunia	14	Seed	Greenhouse	(6)	It is essential to be available during the early stages of development, when the seedlings depend on frequent care and careful control of their surroundings.
Polyanthus	17	Seed	Cold frame	(4)	Modern strains are more temperamental than the old-fashioned ones, so start with the old before trying the new.
Snapdragon	12	Seed	Cold frame	(3)	Another method that is very weather dependent and works best in milder districts.
Wallflower	11	Seed	Outdoors	(2)	Very simple – why buy plants?
Bergenia 'Sunningdale'	40	Cuttings	Cold frame	(4)	One way to increase a stock rapidly. If only a few plants are needed, simple division is easier.
Geranium macrorrhizum	30	Cuttings	Cold frame	(3)	One of the quickest and simplest ways to grow excellent ground-cover to reduce maintenance.
Penstemon 'Firebird'	34	Cuttings	Greenhouse	(5)	Prospects vary depending on winter weather: in severe weather the rooted cuttings may die unless well cared for.
Phlox (border)	26	Cuttings	Cold frame	(3)	Many other herbaceous plants are easily grown by this method. The cuttings form roots quickly, without requiring a great deal of care.
Primula denticulata	42	Cuttings	Cold frame	(5)	These can also be grown from seed, and root cuttings are only used for special forms.
Violas	35	Cuttings	Cold frame	(4)	A rapid and reliable means of increase.
BULBOUS PLANTS					
Daffodils	1	Division	Outdoors	(1)	Makes the most of natural means of increase.
Bluebells	19	Seed	Cold frame	(4)	Not difficult, but care and patience over several years will be needed for success.
Lilies	41	Cuttings	Greenhouse	(5)	Developing bulbs need attention and careful management for two or more years.

Plant	Recipe	Method	Equipment	Skill Level	Comments
CONSERVATORY AND PATIO PLANTS					
Busy Lizzies	43	Seed	Greenhouse	(6)	Looking after young seedlings in late winter requires daily care and good management.
Freesia	50	Seed	Greenhouse	(4)	Not difficult, although neglectful watering in hot summers can lead to failure.
Geranium	45	Seed	Greenhouse	(6)	Skilful management is essential, otherwise the plants will not survive the winter.
Schizanthus	47	Seed	Greenhouse	(7)	The plants make almost all their growth in winter, when things easily go wrong.
Viola 'Prince Henry'	44	Seed	Cold frame	(3)	A straightforward method when the weather is usually on the gardener's side.
Chrysanthemum	48	Cuttings	Greenhouse	(5)	These will do well only with care and attention of the right sort at the right times of year.
Streptocarpus	49	Cuttings	Greenhouse	(7)	A rewarding plant to grow, but one that depends on good conditions and regular care.
Verbena 'Sissinghurst'	46	Cuttings	Greenhouse	(3)	Rapid results and easy to grow. An ideal plant for the beginner.
SHRUBS, CLIMBERS AND TREES					
Bamboo	7	Division	Outdoors	(3)	Should present few problems provided plants are not neglected during the first weeks.
Purple filbert	5	Division	Outdoors	(1)	Reliably successful with minimum demands.
Rhododendron 'Pink Pearl'	6	Division	Outdoors	(3)	Plants can practically be left to look after themselves, but will take their time.
Daphne mezereum	22	Seed	Cold frame	(6)	Care and attention needed at intervals over eighteen months for success.
Lavender 'Munstead'	20	Seed	Cold frame	(4)	Not difficult provided watering can be attended to regularly in hot weather.
Oak tree	23	Seed	Greenhouse	(3)	Straightforward, and a rapid way to produce a well-grown tree.
Rhododendron Hybrids and Species	21	Seed	Greenhouse	(7)	The method involves several stages when more-than-average skill is needed.
Japanese azaleas	32	Cuttings	Greenhouse	(5)	Good management and a certain amount of time at critical stages are needed for success.
Ceanothus	33	Cuttings	Greenhouse	(5)	Provided the right equipment is available and time given for watering, good results follow.
Clematis orientalis	25	Cuttings	Greenhouse	(5)	Daily attention essential for the first week or two, till the cuttings produce roots.
Erica carnea 'Springwood White'	31	Cuttings	Cold frame	(4)	Careful siting of the frame is more important than frequent attention.
Ivy	37	Cuttings	Cold frame	(2)	Tough, hardy cuttings can more or less be left to look after themselves.
Hydrangeas	28	Cuttings	Outdoors	(2)	Very few demands on skill or care.
Lavender 'Old English'	24	Cuttings	Greenhouse	(5)	Needs care and attention for a week or two while the cuttings form roots.
Mock Orange	27	Cuttings	Plastic cloche	(2)	Few problems except in very hot weather.
Roses	36	Cuttings	Outdoors	(2)	Few demands on time or skills.
Roses	39	Cuttings	Cold frame	(5)	Small cuttings need careful management and frequent attention.
Spiraea 'Gold Flame'	29	Cuttings	Cold frame	(4)	Success depends on siting the frame for the right balance between sunshine and humidity.
Thuja 'Rheingold'	38	Cuttings	Cold frame	(4)	Tough, resilient cuttings that will survive some mismanagement.

Contributors to Success

Frames, cloches, greenhouses and other equipment are the subject of the next chapter. But the wellbeing of seedlings and cuttings depends as much on how they are treated as where they are. In other words, whether they are fed and watered to their liking; kept warm enough or cool when they need to be; cosseted in a humid atmosphere or left exposed to the four winds. These contributors to welfare recur in all the recipes – often in terms which are bound to seem imprecise to anyone looking for help. Hopefully the comments which follow will explain or justify some of these imprecisions.

TEMPERATURE

Plants live in a world where temperatures change all the time and, since they are not warm-blooded, their temperatures change too. They vary with the seasons, go up and down as day follows night, and fall or rise in rapid succession as clouds cover the sky or the sun breaks through. Trying to provide precisely controlled temperatures is a waste of time, money and energy when propagating plants, and very often there are other good reasons for not doing so. The first is that it is the surest way to a nervous breakdown. The second is that some of these changes are used by plants to guide their development.

The great majority of the plants in our gardens come from temperate parts of the world and are hardy. They naturally do well in cool conditions and can endure cold, including sub-zero temperatures. They are tolerant of high temperatures, usually responding well when these last just a few hours at a time. On the other hand, prolonged high temperatures may cause them to suffer and can even cause them to die.

Human responses to changes in temperatures, buffered by our constant body temperature, are not a reliable guide to the reactions of plants from temperate parts of the world. The levels which we find most comfortable can be 10°C (18°F) higher than we should aim for when growing garden plants. A very broad view of the significance of temperature from the plant propagator's point of view is shown opposite.

Recommended temperatures appear in many of the recipes, but these should be regarded as guidelines, not as precise instructions. For example, an injunction to adjust a thermostat on soil-heating cables so as to maintain a temperature of 15°C (59°F), does not imply that exactness is crucial; only that temperatures *in the region* of 15°C (59°F) (perhaps plus/minus 5°C) are the aim for as much of the time as possible. Cooler may slow proceedings; much warmer might cause damage.

VENTILATION

Ventilation is essential for the wellbeing of most plants, but the levels needed are very hard to define. Anyone scanning the recipes must be struck by the paradox that gardeners go to great trouble to make cosily airtight, draught-free frames and greenhouses in which to keep their plants warm – then insist on keeping the ventilators open for most of the time!

The recipes more often specify keeping ventilators open than keeping them closed. This is because ventilation is the most effective way to control temperature and humidity, and both rise quickly in enclosed spaces to levels which many plants find excessive. You may respond appreciatively to humid warmth reminiscent of a sultry tropical evening when you walk into your greenhouse, but you can be sure that any

THE EFFECTS OF TEMPERATURE

TEMP	EFFECTS AND SIGNIFICANCE
Less than 0°C (32°F) (very cold)	Temperatures below freezing point can kill any plant. Hardy plants should never be moved straight from high temperatures – say, from a heated greenhouse – to a place where they are liable to experience frost. If kept for a week or so in cool conditions they will 'harden off' and become frost tolerant. All plants have a limit to the degree of frost they can tolerate, and their management must take account of this. Young plants, including recently rooted cuttings, are often less hardy than mature ones.
0 to 5°C (32 to 41°F) (cold)	Most plants make very little or very slow growth, but hardy plants and even those which are slightly tender should suffer no ill effects. This range of temperature will not adversely affect plants overwintering in greenhouses or frames and, when days are short and light is inadequate, may provide more healthy conditions than higher temperatures.
5 to 10°C (41 to 50°F) (chilly)	Many seeds will germinate and the tougher, more resilient cuttings will form roots – but most will do so slowly. The temperatures may be below the optimum for growth but are unlikely to do damage. Night temperatures at these levels, in combination with higher day temperatures, can be very favourable for plant growth and seed germination.
10 to 20°C (50 to 68°F) (cool/mild)	These provide very favourable conditions for growth and development of many plants, and for germination of the seeds of many hardy plants. They can be optimal conditions for root growth. It is a favourable range for the production of roots by cuttings of many kinds of plants. Stress-related problems, such as susceptibility to disease and growth disorders due to physiological upsets, are less likely than at higher temperatures.
20 to 25°C (68 to 77°F) (warm)	These are good conditions for daytime growth and development of cuttings and germination of seeds, but may cause stress-related problems in hardy plants – particularly to roots – if maintained continuously. Plants from sub-tropical and warmer parts of the world are likely to peform well even at continuous high temperatures.
25 to 30°C (77 to 86°F) (hot)	Cuttings and seeds of hardy plants are likely to be stressed at these temperatures if exposed to them for more than a few hours at a time. They should be alleviated by the use of shading, sprays of water and ventilation.
30 to 40°C (86 to 104°F) (very hot)	The upper limit is very close to lethal temperatures. The lower is likely to cause severe stress, growth disorders and extreme susceptibility to disease in hardy plants exposed for more than a short time. They are undesirably high temperatures even for tropical plants, and are likely to reduce performance and increase demands on management.

seedlings and rooted cuttings of temperate perennials and shrubs it contains are suffering in silence.

There are times when plants must be nursed, such as recently potted divisions, newly pricked-out seedlings perhaps, or cuttings freshly taken and still to form roots. Then for a few days, perhaps for a week or two, warmth and humidity are the means of survival. But as soon as that dependent stage is over, they will do better in fresher conditions. Hence the repeated instructions in the recipes to keep the lights of frames raised, except during bitterly cold winds and on nights when damagingly severe frosts are likely.

FEEDING FOR GROWTH

After leaving school I worked for a time at Long Ashton Research Station, feeding plants growing in pure, acid-washed sand with mixtures of chemicals containing different nutrient levels. It taught me how fundamentally plants respond to the nature of their sustenance. Not simply that, up to a point, they do better with more – but also their ability to survive on scarcely any. It is extremely difficult to starve plants to death, though many gardeners – amateurs and professionals – regularly reduce them to shadows of their better selves.

It is often difficult to tell whether plants are doing less well than they should because they are short of nutrients. Experience, skill and resources are needed to make feeding an exact science; in their absence it is better to treat it as an inexact science, and supply nutrients more or less by rule of thumb. Much the most important thing is to supply them!

Plants in containers can be fed by sprinkling fertiliser over the compost, using either crystals, powders or granules. It is safer and easier, and the results can be more immediate, to apply the nutrients in solution, using proprietary products sold as crystals or concentrated solutions. Their strength can be varied very easily depending on how much water is added before use. Feeds should always be applied as balanced, complete nutrients – in other words, they should contain all the major elements needed by plants for healthy growth, in proportions which relate to the overall needs of the plants. But that is a form of words useful for answering questions in exams, which conveys little to someone wondering what to give a frameful of rooted cuttings, obviously in need of something – but what?

My own practice is to use proprietary fertilisers with high levels of potassium, and moderate levels of nitrogen and phosphorus; most also contain other nutrients. Soluble crystals of Phostrogen or concentrated liquids such as Tomorite, are good examples (an organic alternative could be used if preferred), and high-potash liquid feeds are specified throughout this book. Frequently the recipe suggests that the feed should be used at greater concentrations than those recommended by the manufacturer (5× for five times: 1× for the manufacturer's recommendation, etc). This is likely when applying feeds to rooted cuttings growing in a cutting compost with no nutrients at all. Instead of topping up a depleted level – as is the case when used to supplement a potting compost – the application has to supply all that is needed to support healthy growth and development.

USING PESTICIDES AND ROOTING HORMONES

There is one obvious omission from these recipes: they contain little information about chemical control of pests and diseases – no spraying schedules or other routines – designed to preserve the health and life of seedlings or cuttings.

The reasons for this form an almost circular argument. The first is that I seldom find them necessary; the second is that I prefer to avoid using chemicals unless forced to by a problem which cannot be resolved some other way. In the restricted, intimate conditions of a greenhouse used for propagation, most insect pests can be avoided either by doing things they don't care for or, if they do make an attack, by destroying them by hand. Rubbing out aphids is remarkably effective – the survivors get so upset they move off and most lose themselves before they find their way back.

Attacks by fungi, particularly the water-borne pathogens that cause seedlings to damp off, and the stem and root rots which destroy cuttings before they form roots, are serious problems but hard to control with chemicals. These, and the grey moulds that kill stems and foliage of plants in frames and greenhouses, are better controlled by taking care about the ways the plants are treated and the conditions in which they are kept. If they

Hormone rooting powders (and liquids) encourage cuttings to form roots. Most also contain fungicides to protect against fungal infections

become problems there is something wrong somewhere, which should be looked into and put right. Each recipe contains a section headed 'Possible Problems' where more information about avoiding/controlling pests and diseases will be found.

Chemicals that are regularly specified are the hormones that can be applied to cuttings to encourage them to form roots. All the widely available brands are based on the same hormone, formulated either as a powder or a liquid and usually combined with a fungicide such as benomyl or captan, which gives some protection against stem rots. The powder is usually specified in the recipes, but this implies no prejudice against liquid formulations which could be substituted if that is your preference.

Dip ends of cuttings into powder. Tap lightly on edge of lid to dislodge surplus powder

WATERING

In deference to a bad habit, almost universal amongst gardeners, watering has been left till last. This, and feeding, are far and away the most skilled aspects of growing plants and yet it is the norm on many nurseries to find the watering entrusted to the least experienced members of staff. It may even be the schoolboy who turns up part-time on Saturday mornings for pocket money and barely knows a shrub from an alpine.

Over and over again the recipes will suggest that you water 'thoroughly' or 'sparingly' or 'carefully' – as if you would do it any other way! These subjective, imprecise terms must seem very unsatisfactory when the health and life of the plants, and the success or failure of everything the book is about, depends on understanding precisely what is meant. But I have found no way to describe precisely a skill which depends entirely on judgement of particular circumstances – and very often a capacity to take into account what the future holds as well.

Careless watering costs plants. One of the most worthwhile habits any gardener can acquire is the knack of watering carefully and methodically

On the other hand, there are a few golden rules, which help to make watering easier or more effective:

1 Always water newly potted plants and pricked-out seedlings immediately they have been planted.
2 Use a watering can, not a garden hose, with a close-fitting rose on the spout.
3 Use a well-made, well-balanced can that neither leaks nor drips from the seams or the edges of the rose. A galvanised metal can with a long spout is ideal; costly to start with but a wonderful investment that will last for years.
4 Before starting, make sure all the pots are standing level, and in an orderly block, without unnecessary gaps between pots and no stragglers beyond the edges.
5 Decide how many pots, trays etc you want to water from each canful before you start. Vary the number depending on the weather, the size and rate of growth of the plants etc. Never assume that all the plants in a greenhouse or frame will need the same amount. Hold the can with the rose as close to the plants as possible, and go over them several times till it is empty.
6 Start by going all round the edges of the block and then move into the centre. Remember that plants on the edge are likely to dry out more than those in the centre, and more likely not to get their share.
7 Make it a rule to water thoroughly and then leave the plants until they need more. Don't keep them constantly topped-up and soggy with daily sprinklings.
8 When growth is at a low ebb during the winter, reduce watering to the minimum necessary to stop the compost from drying out. Water during the morning on sunny days to allow the foliage time to dry before nightfall.

Water butts are havens for the fungal diseases that kill cuttings and young seedlings. Use tap water for vulnerable plants like these, and reserve stored water for more mature plants that are better able to resist infection

Finally, to be or not to be a sprayer or a soaker? Compost can be wetted in two ways: either sprayed from above using a watering-can, or allowed to soak by standing containers in a tray or bowl of water. I enjoy a long soak in a bath for myself but prefer not to subject my plants or seedling to one. In my view the ideal pattern to follow when watering is a short, sharp, spring shower with well formed, substantial drops of water that run off the leaves as they hit them, and wet the compost quickly. The only time that a soak becomes essential is when peat compost dries out and fails to take up water sprinkled on it from above.

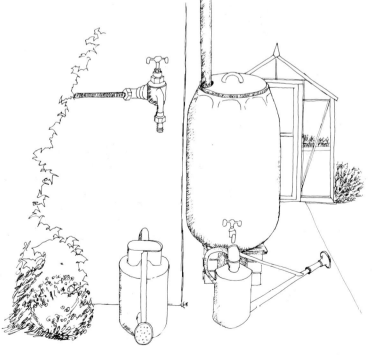

3 EQUIPMENT AND FACILITIES

Investing Sensibly

Many people assume that an ivy is just ivy, a variant of the evergreen climber that appears throughout our woods when the leaves fall off the trees in the autumn. But this is a variegated form of the Turkish ivy *Hedera colchica*, with large leaves and a hardy nature. 'Paddy's Pride' was its name – now we should call it 'Sulphur Heart'. Cuttings in winter (recipe 37) provide a simple, rapid means of propagating one of the brightest and most amenable of winter foliage plants

Gimmicks for Sale

Garden centres dazzle me with displays of things to buy. When I want to sow a few seeds, they will offer me twenty shapes and sizes of pots and seed trays made from a dozen materials. A score of composts, and fifty cunningly formulated fertilisers. Alluringly packaged propagators with electronic controls, priced from the sublime to the ridiculous. Red, blue, green, yellow – and pink if you like – plastic watering-cans (colourful but incapable of sprinkling without dripping). Prepacked, ready-to-assemble cloches, frames and mini-greenhouses of such various sizes, shapes, designs and materials that I am at a loss to know what makes a frame different from a cloche, and either of those different from a mini-greenhouse. And, at the top of the range, conservatories and greenhouses: lean-to, upright, hexagonal, domed and traditionally ridged – glazed with plastic or glazed with glass, till my head spins with the choices and I creep back to my car, glad to retreat to a simpler world.

For those who want to sow seeds, take cuttings or divide plants, there is a simpler world: a world scarcely concerned with spending money on complex equipment. It is a place where tried and proven methods suffice; and doing things when the time is right, in ways that are likely to give good results, are the keys to success.

The recipes in this book have been chosen to introduce you to this world. They will show you, step by step, how to use simple equipment effectively, by becoming familiar with its problems, and profiting by its opportunities. They will show you that it is not necessary to use a great variety of different containers, composts and fertilisers. I would rather vary the numbers of cuttings in a pot than vary the size of the container; use a familiar liquid feed as fertiliser, varying the concentration or frequency of application, rather than pursue fertilisers for courses through the shelves of garden centres. I find it helpful to discover the foibles – good and bad – of a very few potting composts and one or two cutting composts. Then I have some chance of learning how often and how heavily to water at different times of the year, and when feeding will be needed to boost declining food reserves.

Choosing Containers

Lesson number one: don't sow seeds in a seed tray – unless you need to grow five hundred seedling zinnias or fifteen hundred snapdragons, for that is what they can hold.

Lesson number two: don't use rubbish salvaged from refuse bins – eg yogurt pots, take-away dishes, polystyrene cups and the like – in place of well-designed pots and trays.

CONTAINER	ADVANTAGES	DISADVANTAGES
Disposable pots made of peat, paper and other materials that can only be used once.	Readily available: likely to be stocked by almost all suppliers of horticultural sundries.	Much more expensive than re-usable containers, because of the need to purchase new ones every time they are needed.
Round pots made of terracotta, plastic or peat etc.	Traditional shape still favoured by many gardeners, and readily available.	Do not make the most economical use of bench space or when packed into pot carriers.
Square pots made from various kinds of plastic.	Occupy about 30 per cent less space than round pots, and can be packed economically into carriers.	Retail horticultural outlets are unlikely to stock these, but they are widely used commercially.
Seed trays made of plastic or wood.	Economical containers when used to grow young plants. Transmit warmth efficiently on heated benches.	Their large surface areas make these very uneconomic containers when small numbers of seeds or cuttings are needed.

The most practical containers for sowing seeds, holding cuttings and potting up divisions, rooted cuttings and small plants are small 7cm (2½in) square pots made of flexible, long-lasting black plastic. I use these almost exclusively. Their only drawback is the reluctance of retail outlets to stock them, so they can be difficult to buy. But nurseries use them in vast numbers and are often willing to sell once-used containers cheaply.

Containers for plants come in all shapes, sizes and materials. The most economical are those which can be used over and over again. Square pots occupy less space than round ones. Space-gobblers – like seed trays – should never be used for sowing seeds. Multi-celled containers are less flexible and more difficult to manage well than separate units

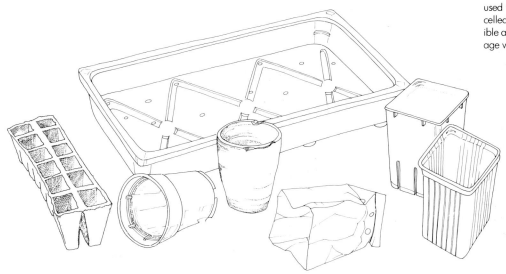

Handling small pots one by one each time they are moved is very tedious, so follow the standard nursery practice of holding them in pot carriers. A seed tray is exactly the right size to hold fifteen 7cm (2½in) square pots, but other flat shallow boxes or trays could be used, including the flat trays in which tomatoes are sold.

Large pots for growing on come in metric sizes from one litre upwards – the size will be stamped on the underside of the pot.

Making the Most of Composts

Compost is an ambivalent term that may refer to the products of the compost heap; to mixtures of loam, grit, peat, fertiliser etc, used to grow plants in pots; or to mixtures of some of these used to support cuttings while they form their roots.

It is possible to make your own potting composts from peat, grit, loam etc, adding chemicals to provide the nutrients that plants must have. There are risks of adding unsuitable fertilisers, overdoing the quantities or not mixing them in thoroughly. These can lead to death or injury of seedlings or young plants, and after various disasters I rely entirely on ready-made proprietary products. Cutting composts are simpler and I always make my own.

The use of peat has become embroiled in controversy because it is obtained from deposits, which are not renewable resources – at least not on the scale involved. Some say that its extraction results in the wanton and needless destruction of places where rare or unusual plants and animals live. The arguments are by no means unchallenged; however, alternatives to peat are available, and it is up to individuals to form their own opinions on the issues involved.

MATERIAL	ADVANTAGES	DISADVANTAGES
Loam Obtained from the top 20cm (8in) or so of permanent or established grassland.	The foundation of John Innes-type composts. At its best, it makes easy to manage, long-lasting composts. Holds nutrients well and takes up water without difficulty.	Seldom at its best. Loam is a very variable material and deteriorates rapidly when packed in plastic bags.
Peat Used as a substitute for loam in a wide variety of composts for the last forty years.	Exceptionally pleasant to handle. Consistently good quality. Long experience of its properties have ironed out most problems.	Not easy to manage, and success depends on skilful watering and careful attention to feeding. Will not re-wet after drying out, and has little hold on nutrients.
Peat substitutes A great number of these are being promoted, including coconut fibre, composted domestic rubbish and waste products of varied origin.	Very often the main advantage put forward by their promoters is that they are not peat! More tests and operating experience are needed to assess their value.	Qualities and limitations are often obscure and unforeseen problems occur. The consistency or quality control of products can vary widely from one bag to another.

MATERIAL	ADVANTAGES	DISADVANTAGES
Grits and sands Obtained from various sources. It may be necessary to distinguish between those derived from basic or acidic rocks*.	Normally used to open out a compost so that water runs through it freely, to prevent water-logging and make it more permeable to air (oxygen).	Hold little or no water and are liable to dry out rapidly. Add considerably to the weight of a compost.
Absorbent minerals These include perlite, vermiculite and calcined clays.	Possess most of the properties of grit with the added advantage that they are water-absorbent and less likely to dry out rapidly.	Supplies can be variable, but when quality and availability are assured these materials have few, if any, disadvantages.

*Some plants, particularly ericaceous plants such as rhododendrons and summer-flowering heaths, may be damaged if grown in composts containing grits derived from basic, lime-rich rocks.

Whatever your feelings about using peat for horticultural purposes, reliable alternatives are not easy to find. Loam-based composts deteriorate so badly when marketed in plastic bags that most emerge in an unusable condition. They are marketed under the tag of 'John Innes' but have little relationship with the composts Mr Lawrence developed while working there. Peat substitutes like coir and other waste products have not yet been thoroughly tested. Whenever I have used them problems have followed, either due to poor quality control or to adverse effects on certain plants. Recipes throughout this book stick to proprietary peat-based composts for potting up young plants and divisions. That is no reason why alternatives should not be tried – with caution.

Cutting composts are a different matter. Peat-based mixtures are less than ideal for use by amateurs, inexperienced in their management. Mixtures of horticultural grit and absorbent minerals such as perlite or vermiculite are much easier to manage, as they drain more freely and take up water without difficulty if they do happen to dry out. I have never found a great difference between these two minerals, but perlite is recommended as a component of cutting composts, and vermiculite is preferred for sowing seeds. The standard cutting compost recommended in almost all the recipes is simply a 50/50 mixture of grit/perlite.

Choosing Equipment

Plants can be propagated without any equipment at all – just the flower beds in your garden – or in laboratories under totally artificial conditions. Between these extremes the options are almost infinite. The guiding principle to remember is that the more elaborate the equipment, the more care, time and effort will be needed to make it work. Hardwood cuttings in rows in the garden can be left to get on with things their own way – at some seasons for months at a time. Cuttings in a greenhouse need to be looked at

A greenhouse is the most valuable asset a propagator can have — but only works well with regular
care and attention.

Always site it close to the house, where it is easy to
reach — even in the dark or in bad weather — and
cheap and simple to bring in electrical and water
supplies. Make a set of small frames close to it in
which to strike cuttings, grow on seedlings or harden
off tender plants, and have a sheltered level area
around it where plants in pots and seedlings in trays,
can be stood out close to a source of water

every day to check whether they need to be watered, fed, heated, ventilated, potted up, damped down and so on.

In many respects well-sited, properly constructed frames provide a compromise that gives good results without too much attention. It is always worth making frames in sets. A pair, or even four small ones are so much more flexible than one large one.

Summary of Equipment that can be used to Propagate Plants

NURSERY BEDS

The plants you buy, even large shrubs and trees, mostly come in containers. They thrived because someone took the trouble to water, feed, repot when necessary, or stand them up after gales, almost every day for months or years. Home-grown plants can be treated in the same way – if you have the time. They will need much less attention planted in a nursery bed while they are small. Similarly, seeds, divisions and some cuttings make fewer demands on time or skill when sown or planted in the open than under cover in frames, greenhouses or propagators.

A small nursery can be made in a corner of the garden, where perennial plants, shrubs or even trees can be grown until they are large enough to hold their own. The place should be sheltered, sunny and well drained. It is usually helpful to plant in raised beds not more than 1.5m (5ft) wide. Everything possible should be done to improve the

Care and skill are needed to grow plants well in containers. Plants in nursery beds are easier to look after and more likely to develop into plants fit for the garden.

The beds are easier to manage if they are raised, and should not be more than 15m (5ft) wide. Top-dressings of grit, leafmould, mulching materials or garden compost improve the soil. Contrary to popular belief, bare-rooted plants usually transplant better than those grown in containers — but most should be moved only between early autumn and later winter

21

existing soil by digging in grit and plenty of organic matter, such as garden compost, well-rotted muck, composted bark etc. A water supply close at hand is an advantage.

MERITS	LIMITATIONS
Plants in a nursery bed more or less look after themselves. They will not depend on the almost daily care required when they are in pots. Most shrubs, and almost all perennials, grow much better in beds than in containers. Plants with free root systems usually transplant with fewer checks than those grown in pots.	The plants are at the mercy of the weather so only those that are totally hardy should be left in the bed during the winter. Plants that are not needed in the garden must be disposed of before they develop such extensive root systems that they become fixtures.

CLOCHES

These can be bought ready-made of glass or several plastics, some much more durable than others. Or they can be set up by stretching a length of bubble polythene over metal half-hoops.

MERITS	LIMITATIONS
Cloches provide a cheap, flexible and effective way to protect plants, cuttings and seedlings. In the winter their main value is the protection they give from wind chill. In summer they can be used to provide the humid atmosphere essential for the survival of cuttings.	Some types are unstable and too easily displaced by gales, or are vulnerable to damage during daily use. Clear film is liable to cause over-heating during sunny weather. Milky white or, preferably, bubble polythene maintain more satisfactory conditions inside the cloches.

SEED TRAYS WITH PROPAGATOR TOPS

Many manufacturers market simple assemblies consisting of a plastic cover that fits over a seed tray. Some form of semi-controllable ventilation is usually built into the top.

MERITS	LIMITATIONS
These are cheap and effective substitutes for a plastic bag tied over a pot of cuttings. They provide an economical unit for anyone who needs to produce small numbers of plants. They can be managed in ways which make very small demands on time or energy.	Tops are likely to be made from brittle, easily broken plastic. Enclosed space within the propagator tops overheats during short periods exposed to direct sunlight, and careful siting is important. Bubble polythene can be used to make a light-transmitting, heat-shading canopy.

COLD FRAMES

These are far and away the most economical and useful facility for propagating plants. Ready-made products are usually relatively expensive and often more suitable for growing plants which need high light intensities – such as lettuces and cucumbers – than for use as propagating frames. Home-made frames can be made simply and economically out of wood and bubble polythene to fit whatever space is available. The good insulation provided by both, and the light-diffusing properties of the polythene, make them ideal materials for the job.

A frame can be as productive as an unheated greenhouse, at far less cost. The difference is that you know what the greenhouse feels like every time you walk into it, while you have to imagine what the interior of the frame is like; consequently you need less experience and less imagination to manage things successfully in a greenhouse.

MERITS

The most economical way to provide easily managed, flexible propagating space. Pairs or sets of frames provide more flexibility than single units, allowing for the needs of cuttings, seeds or divisions with different requirements or at different stages of development, when some need more ventilation than others for healthy development. An effective, low-cost way to protect young hardy/semi-hardy plants from wet and the extremes of cold during the winter.

LIMITATIONS

Framelights covered with glass are awkward to handle, liable to produce high temperatures when in direct sunlight and can be dangerous. Bubble polythene is a safer, lighter material that maintains almost ideal conditions within the frame. A frame provides limited protection from winter cold and is not suitable for tender plants. Plants are vulnerable to attack by slugs and mice, and cats may be tempted to make use of the frames when the ground outside is frozen.

Many plants can be propagated as easily in frames and cloches as in a greenhouse, but at a fraction of the cost.

Several small frames are easier to use, and more flexible, than one large one. Timber sides and lights covered with bubble polythene are cheap, effective and robust. Plastic cloches are easy to make from lengths of metal rod or heavy-gauge wire covered with bubble polythene

OUTDOOR HEATED FRAMES

The traditional outdoor frame heated with hot waterpipes from a boiler, or set up over a hotbed, is almost extinct. But nowadays they can be made without much difficulty, using soil-heating cables controlled through a thermostat. Several companies market a variety of upright plastic-covered structures – usually referred to as though they were substitutes for greenhouses but in reality variations on a frame.

MERITS

Used skilfully and imaginatively these provide an economical alternative to a heated greenhouse. They are particularly useful in spring and autumn, and in mild districts, for maintaining frost-free conditions if night-time temperatures fall only a few degrees below freezing point.

LIMITATIONS

Severe conditions experienced during mid-winter will result in temperatures below freezing point in outdoor frames. Great care must be taken to ensure that the frames do not dry out unnoticed whenever the heating cables are switched on.

Even a small greenhouse can be used to propagate thousands of plants every year. There is no need to spend a fortune keeping the space inside it warm during the winter: soil-heating cables buried in the benches provide the protection from frost that young cuttings and seedlings need at little cost. During very cold weather a blanket of bubble polythene laid directly over the plants traps warm air in a sandwich around them

UNHEATED GREENHOUSES

Many greenhouses have no heating of any kind and are scarcely more productive than frames. For this reason, it is a waste of money to spend the amount needed to put up a greenhouse, unless you then install an effective method of keeping plants within it warm.

HEATED GREENHOUSES

Greenhouses can be bought in many shapes and sizes with metal, timber, plastic or concrete structures glazed with glass or transparent plastics. They are always expensive but the price varies greatly depending on size, design and materials used. A propagating house can pay for itself within a year, and the design and the materials it is made from play little part in its success. Even the smallest can be used to produce enough plants to fulfil the needs of all but the most ambitious amateur gardener – we are talking here in terms of thousands of plants a year. If a greenhouse is needed only to propagate plants, the answer is to buy the cheapest possible design of one of the smallest models, constructed from the least expensive materials. Spending money on appearance or other attributes makes sense only if other factors have to be considered.

Manufacturers are reluctant to provide their greenhouses with enough ventilators. Even the smallest must have a vent on either side of the ridge. And do not listen to the proposition that doors can be left open to provide ventilation: open doors are invitations to cats, and other marauders, who will enter and bask amongst your seedlings.

MERITS

Provides a flexible, protected environment in which to propagate plants and grow them on. Combines a pleasant atmosphere in which to work with very favourable conditions for growing a wide variety of plants. Even a small house has great productive potential.

LIMITATIONS

A very expensive facility and one that can be justified only if well used. Requires almost daily attention/supervision throughout the year to obtain good results. Good ventilation is essential and must be provided for when buying.

HEATED BENCH IN GREENHOUSE

Greenhouses can be heated in a variety of ways. It is extravagant and unnecessary to heat all the space inside a house used to propagate plants, and economical and effective to heat one or more of the benches with soil-heating cables controlled through a thermostat.

MERITS

Heated surface is close to the plants, where it acts most effectively and economically. Control through a thermostat largely eliminates waste. Very effective when used to heat a propagating frame constructed on the bench, or when combined with a blanket of bubble polythene laid over the tops of the plants during cold nights.

LIMITATIONS

The rate of heat input is low and may not be able to keep up when outside temperatures fall rapidly. There is little residual heat reserve and failure of the electrical supply results in very rapid loss of heat.

HEATED PROPAGATORS

These are marketed in various sizes and degrees of complexity, from inexpensive models based on a single seed tray to units three or four times the size, with more elabo-

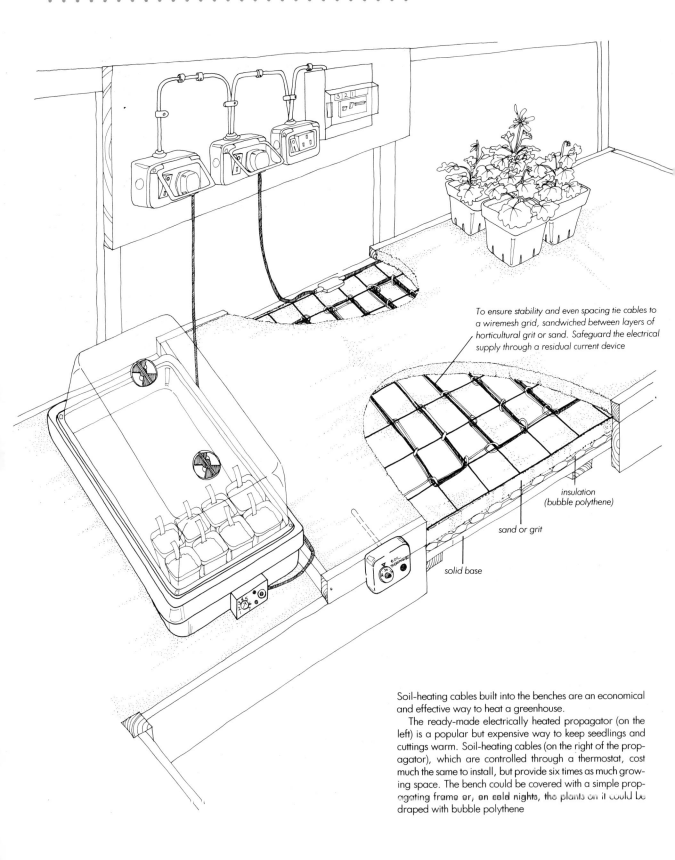

To ensure stability and even spacing tie cables to a wiremesh grid, sandwiched between layers of horticultural grit or sand. Safeguard the electrical supply through a residual current device

insulation
(bubble polythene)

sand or grit

solid base

Soil-heating cables built into the benches are an economical and effective way to heat a greenhouse.

The ready-made electrically heated propagator (on the left) is a popular but expensive way to keep seedlings and cuttings warm. Soil-heating cables (on the right of the propagator), which are controlled through a thermostat, cost much the same to install, but provide six times as much growing space. The bench could be covered with a simple propagating frame or, on cold nights, the plants on it could be draped with bubble polythene rate controls at higher prices. Their manufacturers promote them as 'mini-greenhouses', but this is sales talk. They fulfil few of the functions of a greenhouse, more those of eccentric, prefabricated heated frames.

MERITS	LIMITATIONS
Provide instant, plug-in heated space in which to sow seeds, protect cuttings etc.	Difficult to site as they need to be well illuminated but out of direct sunlight. Do not provide good conditions for healthy plant growth and are a temporary provision that requires very careful management. An extremely expensive alternative to home-made heated frames put together from readily obtainable components.

MIST PROPAGATION UNITS

Success with cuttings depends on keeping them functioning as normally as possible while they make the growth needed for the development of roots. Best results follow when the cuttings are kept more or less at natural light intensities, in an atmosphere almost saturated with water. Mist units make this possible by automatically spraying cuttings set out on an open bench in a greenhouse, whenever they begin to dry out.

MERITS	LIMITATIONS
The favourable environment provided by a well-lit, open bench and constant high humidity makes success possible, even with difficult cuttings. When used with heating cables below the surface, the time needed to complete propagation cycles can be greatly reduced.	Cost can only be justified when large numbers of difficult cuttings are to be propagated. Mist units only work well when carefully maintained, and are not fail-safe. When things go wrong disasters follow, due either to waterlogging or dessication when mist units operate too frequently or not at all.

THE USE OF ELECTRICITY IN PROPAGATION

Note: Electricity is the main source of power used for heating greenhouses, and operating equipment used when propagating plants. It is convenient to use, extremely controllable and need not be expensive, particularly if off-peak tariffs are taken into account. Electrical supplies to greenhouses, heated frames and propagators must always be professionally installed, and be provided with safeguards that prevent serious injury if equipment or parts of the installation develop faults.

There is seldom only one way to propagate a plant, and substitutes can often be found for the equipment used in these recipes, with equally good results. In some cases these have been suggested. However, when I first thought about a book of recipes for plant propagation it seemed essential to present each recipe as clearly and unambiguously as possible. This meant the ommission of ifs and buts and either/or, except in exceptional instances where they begged to be included.

The most important point to remember when using and buying equipment is to start with the simple and build up to the complex. A heated greenhouse, unsupported by frames and space to stand plants out, is a barely manageable white elephant. As an addition to an already flourishing set-up, it becomes a valuable asset.

4 INTRODUCTION TO DIVISION

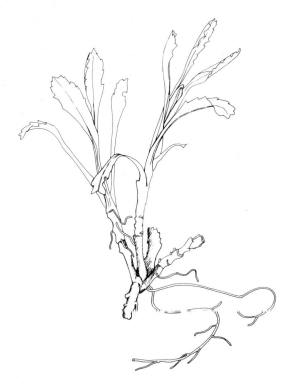

This is my fifth garden. Before moving here, after selling Oldfield Nurseries, I made up my mind that this time there would be no long drawn-out hassle, accumulating plants year by year, until at last the garden was full. I would bring my own supporter's club with me.

The move was to be in the autumn, and during the summer my spare time was spent digging up and dividing some of the herbaceous plants in my old garden. Well-established clumps could be split up to make sixty or more small ones so there was no need to vandalise the place to get all I needed. The divisions, often no more than a single crown, were potted in small square pots, packed forty at a time into plastic carrying trays. They were cosseted for two or three weeks under a framelight in a greenhouse, until roots grew and new leaves appeared, then stood outside.

By the time moving day arrived, I had pulled more than two hundred different kinds to bits, and about twelve thousand plants were ready to go. The people who shifted the furniture declined responsibility for trays of plants, so I hired a large double decker cattle truck and, on a pouring wet day they were loaded up and carted off to Shropshire. Much of the following spring and summer was spent planting them, and by the end of the year they were well established and doing well in their new surroundings.

Many herbaceous perennials – like this Shasta daisy – can be multiplied by division. The mats of shoots and roots are dug up and then pulled apart, or cut when necessary, into small clusters of crowns, or even individual crowns. Note that a crown is a single shoot complete with roots

Dividing plants is fast, effective and certain. It is also one of the ways that plants reproduce naturally. Seeds provide a means of adjusting to new situations, surviving unusual conditions and, ultimately, adapting to changing worlds. But plants also recognise the value of proven worth. Those that have successfully found a place to grow can avoid the uncertainties of the sexual lottery by sticking to a successful formula. Quite simply they divide and rule, producing identical counterparts of themselves until some change in their surroundings, or debilitation from disease, forces a return to sexual dependence.

A hundred michaelmas daises dotting a thousand square metres of a meadow may be the scattered fragments of a single individual. A seedling, settling in amongst the grasses years earlier, grew in annually extending arcs of crowns. The central ones

28

died of age and starvation, but those on the periphery became separated, continued to grow and repeated the process till all signs of their common origin were lost. The streamsides and wooded valleys where I live are white with snowdrops every winter. They appear to be wild, but those who have thought about these things say they are probably descendants of cottagers' clumps. Dug up by moles, or washed away and deposited again and again by floods after heavy rain, they have infiltrated the woodlands for miles. Dogwood branches growing in a hedge are pressed to the ground by snow. Roots develop and they grow into new shrubs. In turn these move outwards, till the formal row along the hedgeline becomes a broad thicket that shelters seedling trees and becomes a little copse.

Gardeners imitate this last process when they produce shrubs from layers – the name given to a branch bent down and held firmly in contact with the soil till roots appear. The connection with the parent plant can be severed once these are well-enough established to support the tops. A month or two later a ready-made, self-supporting shrub can be dug up and planted in the garden.

Almost any branch pressed down to the soil could develop roots. Not surprisingly, they are produced more rapidly and more certainly from young shoots than old ones. The best bet of all is a young branch growing on a young plant, and this is an occasion – the recipes will mention several others – when I will buy a plant from a garden centre or nursery from which to propagate more.

The prospect of filling a garden with stylish shrubs like rhododendrons, witch hazels or magnolias can raise exciting visions. First reactions to the prices on their tags may dispel such dreams; the charms of bargain-basement forsythias might even become irresistible! And how you would regret that decision later, as you imagined how the garden might have looked, if only you had splashed out. But use each plant to produce two or three more within a couple of years and their prices seem less daunting.

Division is easy. So easy that it is often taken too much for granted. Offshoots are planted straight back into the garden, and left more casually than they should be to fend for themselves. You will get away with it when the weather is kind, but frosts during the winter lift up and kill recently divided herbaceous plants; alpines pulled to pieces and dabbed back in are shrivelled by summer droughts. Even the layers from shrubs, which appear to have plenty of roots, need time to readjust and are vulnerable to bad weather. I hate to lose plants like this and, more often than not, it pays to pot divisions up so that they can be given a little more care while their roots and shoots make a new start.

This extra care pays several dividends: more of the plants survive, and you can make more divisions because smaller bits have a chance. Once safely in a container, there is no rush to plant them, and I can spend time deciding where I would like to put them in the garden.

Not surprisingly, some of the recipes that follow reflect this preference and the plants spend a short time in pots. But friends frequently tell me that this is a waste of effort and remind me that one of the great advantages of division is that plants *can be dug up from the garden, pulled to bits and replanted at once*, needing little time and no equipment. If you happen to be busy with other things, and have no wish or time to get involved with potting plants up, watering them and looking after them that is a tremendous bonus. Whether you divide and pot, or divide and plant back into the garden must depend on what suits you. Either way, you will discover that pulling plants to pieces is the most economical and certain way to fill the spaces.

Division is the easiest way to multiply many herbaceous plants. An unusual Shasta daisy – like the shaggy-headed 'Aglaia' – can be propagated with no more difficulty (recipes 2 or 4) than the cultivars 'Esther Read' or 'Wirral Supreme' which are sold everywhere. The beautiful white flowers last well in water, and florist's dyes can be used to change their colour to suit the decor. Only an expert would suspect the deceit

5 MULTIPLICATION BY DIVISION

The Simplest Way to Fill a Garden

 Sooner or later we all fall into the trap of valuing a plant just because it is hard to grow or difficult to propagate. In a more practical mood, I come down to earth and give thanks for all the easy, undemanding plants that are the backbone of my garden.

A great many of these have been produced by dividing up clumps of older plants, with little effort and a very high certainty of success. At its simplest, division is no more than replanting bits of plants separated from their fellows while tidying up a corner. When gardeners were plentiful and herbaceous borders in vogue, this was done during winter refurbishments by digging up old-established plants, splitting the mats of crowns into quarters, eighths or tenths, and replanting them in the newly forked and dunged borders. You can still do it that way, but some of the more attractive perennials resent such summary treatment, and respond better to other methods.

I find that I do less and less gardening of any kind during the winter, and most of the plants that I divide are dug up during late summer. This may seem to be a risky and terrible time to upset a plant by pulling it to pieces, but because it is growing actively it will recover quickly from the shock, and then it is soon ready to go back in the garden. Over the last ten years the results have persuaded me that there is no better season, especially for more delicate plants. In late summer when the soil and the air are warm, and the potting shed a welcoming, shaded place in which to work, it becomes a pleasant, relaxing job – so much more enticing that the prospect of pulling frosted, muddy plants to pieces in a dank, cold shed in wintertime!

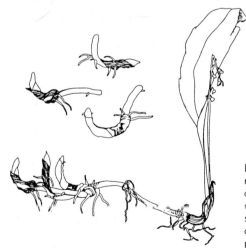

Lilies of the Valley produce mats of root-like stems known as rhizomes. These can be dug up in late summer, cut into short lengths each with a bud at its tip, and replanted in prepared fresh ground

Renovating and Reviving Daffodils   *RECIPE 1*

LIFT/SPLIT/REPLANT: Mid-summer
IN FLOWER: Mid-spring

A very early attempt to produce my own plants came soon after I moved into my first garden, when I dug up some ancient clumps of daffodil bulbs. Crowded together, with one tiny bulb piled upon another after years of division and sub-division, they had almost given up flowering. I lined them out beside the vegetables. The following year a few flowered, the next year many more, and by the third year I was delighted with ninety or more flowering bulbs from each of the original clumps.

——— WHAT TO DO ———

1 Wait until the leaves start to go yellow.
2 Use a fork to dig up clumps of bulbs.
3 Pull the clumps apart, free them from soil and separate each bulb.
4 Pull the leaves off above the top of each bulb.
5 Replant them where you want them to flower. The larger ones will flower the following year; the smaller ones after a year or two.

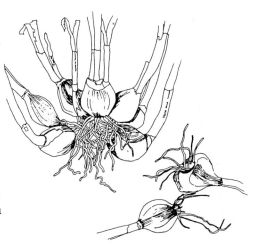

Daffodil bulbs multiply naturally by forming offsets. In time, congested clusters of bulbs develop — very few of which grow large enough to flower. The crowded bulbs should be dug up, pulled apart and replanted individually. They soon grow large enough to produce flowers

POSSIBLE PROBLEMS

SYMPTOMS	CAUSE	REMEDY
Some bulbs hollow and squashy.	Narcissus fly (a)	Burn all the affected bulbs.
Almost all bulbs very small with few, if any, flowers.	Can be the result of starvation due to overcrowding; may be infection with virus.	Will right itself after the bulbs have been separated and replanted in fresh soil. If only a few bulbs start to flower after two or three years, dig them up and burn them
Distorted leaves and flowers. Soft bulbs; raised spots on leaves.	Stem eelworm (a)	Burn every bulb in infested clumps. Plant replacements elsewhere.

(a) The bulbs referred to in the list will not suffer from these, but they may suffer from other pests and diseases.

OTHER BULBOUS PLANTS WHICH CAN BE DIVIDED IN MUCH THE SAME WAY

Bluebell
Chionodoxa
Crocus
Grape hyacinth
Iris
Lily
Lily of the Valley
Montbretia
Ornamental allium
Solomon's seal
Snowdrop*
Snowflake
Schizostylis
Scilla
Tulip
Winter aconite*
*It is better to divide these while they are still in flower

| RECIPE 2 | **Wintry Break-ups of Michaelmas Daisies** |

> LIFT/SPLIT/REPLANT: Mid-autumn to mid-winter
> FIRM IN: Early spring
> IN FLOWER: Early autumn

Michaelmas daisies are attractive, easily grown plants, useful for a bit of colour after other herbaceous plants have faded. The best flowers are produced by young plants, and it is worth taking the trouble to divide them every few years to keep them in good fettle.

WHAT TO DO

1 When the weather is mild in winter, fork out old-established clumps of michaelmas daisies.

2 Break them up into clusters of three or four crowns. (When places for them are not quite ready, pack the divisions in boxes with peat – tomato trays are ideal; put them in a sheltered corner; cover them with 5cm (2in) of straw).

3 Prepare planting positions by top dressing with a general fertiliser at 100g/sq m (3½oz/sq yd). Cover with 5cm (2in) of compost, old manure, composted bark etc, and fork it into the top 15cm (6in).

4 Plant the divisions in the prepared ground, making sure that the crowns are neither more deeply buried, nor closer to the surface than they were originally.

POSSIBLE PROBLEMS

SYMPTOMS	CAUSE	REMEDY
Crowns closely matted together, small and hard to separate.	Crowns in the centre of large clumps are likely to be old and starved of nutrients.	Take the crowns from round the edges of the clumps; throw away those in the centre.
Young plants establish badly; grow poorly or wither away.	Lifted from the ground by frost or not planted firmly enough.	Check crowns during early spring. Firm them in and replant any that are lying on the surface. Water them if necessary.

Many herbaceous perennials can be divided during the winter, but cold, wet winters can lead to heavy losses of less hardy varieties.

HERBACEOUS PERENNIALS WHICH ARE TOUGH AND LIKELY TO SURVIVE BAD CONDITIONS				
Balm	Geranium	Hart's tongue fern	Mint	Sedum
Bergenia	Goat's beard	Helenium	Monkshood	Shasta daisy
Bluebottle	Golden rod	Inula	Obedient plant	Soapwort
Chinese lantern	Globe flower	Knotweed	Perennial	Spiderwort
Comfrey	Globe thistle	Lythrum	sunflower	Yellow
Day lily	Hard shield fern	Meadowsweet	Plume poppy	loosestrife

Delphiniums can be Divided in the Spring   *RECIPE 3*

> LIFT/SPLIT/REPLANT: Early spring
> IN FLOWER: Early summer to early autumn

Delphiniums, like many herbaceous plants, are vulnerable to frosts and long wet spells while recovering from being divided. Then you will discover little black slugs that were attracted by the damaged tissues and, their appetites wetted, developed a taste for healthy shoots too. I prefer to divide these plants when the worst of the winter is over. They need a little more care than the tough plants that can tolerate being dug up and torn apart in mid-winter. Delphiniums start to grow early in the spring and young shoots on barely established plants are susceptible to drought and cold. A few precautions will be needed until new roots get going and the plants are ready to grow vigorously once again.

Astilbe	Dicentra	Heuchera	Monkey flower	Royal fern
Bergamot	Erigeron	Lady fern	Ostrich plume	Rudbeckia
Campanula	Evening primrose	Ligularia	fern	Sidalcea
Catmint	Geum.	Male fern	Pyrethrum	Soft shield fern
Coreopsis	Hellebore	Masterwort	Red hot poker	Veronica

PLANTS WHICH ARE BETTER LEFT TILL SPRING BEFORE BEING DIVIDED

Established delphinium plants can be rejuvenated by digging them up in the spring when new shoots emerge, dividing them into small clusters of crowns and replanting them in prepared soil

33

────────── **WHAT TO DO** ──────────

1 First, prepare places in which to plant the divisions. Top dress with 100g/sq m (3½oz/sq yd) of general fertiliser and cover with mature compost or soil conditioner. Fork lightly into top 15cm (6in) and rake over the surface.

2 Dig up well-established plants and divide them into clusters of three or more crowns.

3 Replant them immediately in the prepared place, making sure they are firm and that their roots and shoots are at the same level as before.

4 Lightly hoe the ground between the plants.

5 Mulch between the plants with 5cm (2in) of composted or chipped bark, or other weed-excluding material.

P O S S I B L E   P R O B L E M S

SYMPTOMS	CAUSE	REMEDY
Young shoots wither in the ground; plants fail to develop or remain stunted.	Lack of water: due to drought or drying winds before the roots have had time to establish.	Water well during dry periods. Consider a covering with plastic mulch during long spells of cold east winds.

RECIPE 4 — Late Summer Divisions of Hostas

```
LIFT/SPLIT/POT: Mid- to late summer
PLANT: Mid-autumn to spring
IN FLOWER: Mid- to late-summer
```

This is the time of year when I most enjoy dividing plants. They are pleasant to handle – no mud! – and each clump can be split into a great many small plants. I prefer to pot them up rather than replant them straight into the garden. The results are better and the young plants can be kept safely somewhere out of the way until I am ready to plant them. No need to pop them in wherever there is space and then have to move them when a better idea comes to mind.

A great many herbaceous perennials respond well to this treatment, amongst them the plantain lilies or hostas. I can always find space for more of these; their attractive green, blue or variegated leaves lighten up any shady place, and their cool pale purple flowers – damned with faint praise by some – fit very well into a woodland setting on hot summer days.

────────── **WHAT TO DO** ──────────

1 Dig up well-established plants with a fork. If they have been in the ground some time, you will need strength, determination and stout prongs to shift them.

2 Divide the clumps into as many single crowns as possible. Start by pulling and twisting, then use a knife to cut through the last links. Old clumps may have to be chopped apart.

3 Cut off all but the youngest leaf and trim the old roots back to about 5cm

34

(2in). Avoid damaging any of the short white roots growing close to the crown.

4 Pot each crown into a 7cm (2½in) square plastic pot, using a general-purpose potting compost.

5 Pack the pots into seed trays or shallow boxes in a cold frame. Water thoroughly – if using a peat-based compost, check that it is wet through.

6 Put a framelight covered with bubble polythene over the plants. Wedge it open about 10cm (4in) for ventilation. In hot, very sunny weather, break the force of the sun by draping a second sheet of bubble polythene or a milky-white plastic fertiliser bag over the frame.

7 Water when necessary to prevent the compost from drying out while the plants establish themselves. This should seldom need to be done more than once a week. After two weeks raise the front of the framelights to about 30cm (12in).

8 When the plants show signs of growth, remove the framelight. The divisions will be ready to plant in the garden after a month, or they can be kept in their pots through the winter.

9 Young plants in small pots will survive severe winters better in cold frames, protected with framelights overhead. Keep the fronts raised except during severe frosts or when cold winds blow.

The best time to divide most herbaceous plants is in late summer

Hostas produce new roots as summer comes to an end. Even though they are in full leaf they can be dug up, separated into individual crowns and replanted. Developing new roots ensure that they re-establish quickly and grow into strong young plants before the winter

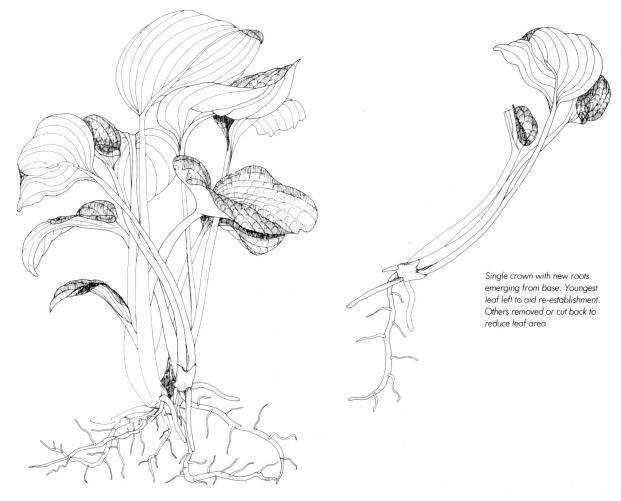

Single crown with new roots emerging from base. Youngest leaf left to aid re-establishment. Others removed or cut back to reduce leaf area

Trumpet gentians are not hard to grow, but, for reasons nobody has fathomed, often produce few flowers. However they need not be difficult to propagate. They produce plenty of seeds, which can be sown in a plastic box (recipe 18), but this is an uncertain method that works well only when skilfully managed. It is simpler to dig up established plants in late summer (recipe 4), separate the individual crowns and pot them up while they produce new roots

. . . .

POSSIBLE PROBLEMS

SYMPTOMS	CAUSE	REMEDY
Divisions fail to establish themselves in their pots.	Either due to overwatering and rotting off, or inadequate watering, leading to dry compost and death from dessication.	Thoroughly soak divisions when they are potted up, then adjust ventilation and shading on the frame according to the weather. It should be unnecessary to water them more often than once a week.
Young plants die during the winter.	Almost always due to the plants freezing while they are sodden wet.	Protect from rain with a frame-light, right through the winter. Ventilate night and day except in severe frost; water sparingly.

OTHER PERENNIALS WHICH CAN BE DIVIDED DURING LATE SUMMER

The list includes many plants which have appeared in previous recipes. When there is a choice, this recipe is recommended.

Achillea	Daisy	Goat's beard	London pride	Red hot poker
Aster	Day lily	Golden rod	Lungwort	Sedum
Astilbe	Dicentra	Hawkweed	Lysimachia	Shasta daisy
Bergamot	Duchesnea	Helenium	Marjoram	Sidalcea
Bergenia	Epimedium	Hellebore	Masterwort	Sisyrinchium
Black-eyed Susan	Erigeron	Heuchera	Meadowsweet	Spiderwort
Brunnera	Eryngium	Iris	Monkshood	Tellima
Bugle	Gentian	Kingcup	Navelwort	Thyme
Campanula	Geranium	Knapweed	Obedient plant	Tiarella
Cerastium	Geum	Knotweed	Peony	Veronica
Comfrey	Globe flower	Leopard's bane	Primrose	Wild strawberry
Coreopsis	Globe thistle	Ligularia	Pyrethrum	

SUCKERS:
READY-MADE SHRUBS

Few of us emerging wounded and bloody from frustrating, unsuccessful forays against suckers amongst our roses would respond warmly to being told that suckers might even be useful. But roses, apples, plums and pears – sometimes viburnums and lilacs too – are not entirely what they seem. These are composite plants in which the variety we know and value has been grafted or budded on to another's roots. This is done for a variety of reasons, some good and some less commendable. Whatever the reason, any suckers produced by the roots of these plants will lack the qualities which we look for in the parts that grow above the ground.

These grafted shrubs and trees are a minority among the plants in our gardens. The majority of the occupants will have been grown from seeds or cuttings or from divisions. They are the same from the tips of their roots to the tops of their shoots, and the suckers they produce grow up to be replicas of their parents. Going round my own garden – or a friend's who has things I covet – I keep an eye open for suckers like these, because they are the easiest possible way to obtain ready-made copies of the plant they come from.

Suckers upset many gardeners by their inclination to appear uninvited where they are not expected. Offspring of stag's horn sumachs erupting around their parents make an impression that decides many people not to let this wonderful shrub anywhere near their gardens. But don't fall into the trap of condemning everything that suckers as a wanton spreader. Some most attractive shrubs, quite difficult to propagate in other ways, produce suckers. Not in great numbers, but perhaps one or two from time to time. These are unlikely to upset anybody and when a plant or two is needed to fill a space or complete a planting scheme, they can be just what's wanted.

The purple-leaved filbert (*Corylus maxima* 'Purpurea') is one such shrub. Well-established plants will send up a few suckers close to their parents, providing a simple way to propagate a plant that is very hard to grow from cuttings.

Stag's horn sumachs are amongst the most versatile and beautiful of all small trees. Yet many people are so disturbed by their suckers that they would never dream of planting one in their garden. The answer is to grow them in situations where injuries to their roots – which lead to excessive sucker production – can be avoided, and to regard any suckers that do appear as a welcome source of free plants (recipe 5)

RECIPE 5 — Suckers from the Purple Filbert

Garden roses are likely to be budded onto rootstocks, and any suckers they produce are briars of little garden value. A few roses are grown on their own roots, including many species, and can be propagated by digging up and replanting their suckers

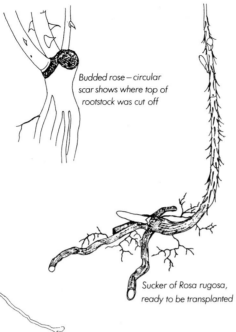

Budded rose – circular scar shows where top of rootstock was cut off

Sucker of Rosa rugosa, ready to be transplanted

WHAT TO DO

1 Compare the leaves, bark and growth of the sucker with young shoots of the shrub. Make sure they are similar, and that the shrub has not been grafted onto alien roots.

2 Use a fork to dig into the ground around the sucker, and expose the roots beneath it. Early winter is the preferred season, but suckers can be moved successfully at most times of the year. If a friend is kind enough to offer you one from the base of a shrub you have admired, be prepared for immediate action.

3 Remove the sucker by cutting the main root on either side, taking with it as many small, fibrous roots as possible.

4 Pot it up into a container just large enough to hold it without bending or cramping the roots, using a standard potting compost.

5 Put the container in a sheltered corner and water thoroughly. Once new roots have developed the sucker can be planted in its intended position in the garden.

Yuccas produce odd-looking subterranean growths known as 'knees'.

Knees are underground suckers and provide a means of propagation. After cutting them off, pot them individually and put them in a cold frame or greenhouse, watering sparingly till shoots appear

POSSIBLE PROBLEMS

SYMPTOMS	CAUSE	REMEDY
Suckers fail to grow, leaves wither.	Not enough roots to support the growth and development of the shoots.	Suckers lifted with few fibrous, feeding roots need careful attention until more develop. Cut shoots back to about 15cm (6in) and keep shaded and well watered.

OTHER SHRUBS WHICH PRODUCE SUCKERS, OCCASIONALLY OR REGULARLY				
Berberis	Elaeagnus	Kerria	Rubus	Snowy mespilus
Buckeye	Exochorda	Lilac *	Russian sage	Snowberry
Butcher's broom	Gaultheria	Mock orange	Sumach	Spiraea
Clerodendron	Hydrangea	Passion flower	Shrubby	Sweet box
Deutzia	Hypericum	Pernettya	honeysuckle	Yucca
Dogwood	Japanese quince	Rose *		

*Check very carefully to make sure these are on their own roots.

LAYING-DOWN SHRUBS: GARDEN SPACE FILLERS

One of the most intriguing houses I never bought was in Sunningdale; a cottage built for a gamekeeper, reeking of tragedy beside a silt-filled lake. I viewed it on a grey November morning: a sad place enveloped in *Rhododendron ponticum*. Planted years before as cover for pheasants, the intertwining stems looped to the ground and rooted where they touched. Their tangled branches formed a joyless, impenetrable, evergreen forest that isolated the cottage from the surrounding commuter-land. It failed to reach its reserve at auction, and the owner refused to treat.

Rhododendron 'Pink Pearl' from Layers — RECIPE 6

The more showy (ostentatious?) rhododendrons are not easy to grow from cuttings: special skills and equipment are needed, which are neither common nor (modest) garden. Yet, like those ponticums in Surrey, large-flowered hybrids like 'Pink Pearl' will produce roots wherever their branches touch the ground, and, once rooted, can be dug up and planted in the garden.

This recipe uses a shrub bought to be propagated. Plants growing in the garden could be used if their young shoots are low enough to be layered. Sometimes these will be produced after cutting off one or two of the older branches close to ground level. But beware. If the plant has been grafted on to a rootstock, the shoots that grow may all come from the stock.

———— WHAT TO DO ————

1 Buy a plant, before mid-winter, with young shoots long enough and flexible enough to be bent down to the ground and up again.

2 Plant 5cm (2in) deeper than usual in well-tilled, fertile garden soil. If necessary, plant at an angle to bring a suitable branch closer to soil level.

3 Mix soil and grit 50/50, and spread a layer 2cm (¾in) deep over the ground around the shrub.

4 Bend the stem to establish where it will touch the ground; slice a sliver of bark 7cm (2½in) long off the lower side with a sharp knife.

5 Push in a bamboo stake so that, after bending the shoot to the ground, the end can be brought up and tied securely to the stake.

6 Cut the bamboo off just below the tip of the shoot.

7 Put a dollop of the soil/grit mixture 4cm (1½in) deep over the stem where it lies on the ground.

8 Leave until the following autumn, watering from time to time in dry summers, then pull gently at the layer to test for roots.

9 If roots have been formed, use secateurs to cut the stem joining the

Note that the shrub shown here has been planted at an angle in preparation for layering one or two branches

Making a layer is the simplest way to propagate one of the large-flowered rhododendrons.

This is done by bending a side shoot to the ground, and holding it firmly in place until it forms roots. It can then be cut free from the parent plant, dug up and planted straight into the garden

layer to its parent, but do not disturb it otherwise.

10 Early the following spring dig up the layer and plant it in the garden. Prepare planting places carefully with plenty of peat or composted bark dug into the top 15cm (6in).

11 Stake firmly, mulch around the plants with a 4cm (1½in) layer of bark chips. Water during dry periods throughout the first summer.

POSSIBLE PROBLEMS

SYMPTOMS	CAUSE	REMEDY
Layers fail to form roots during the first year.	Some species or cultivars form roots less readily than others, or do so more slowly.	Leave them for another year; then look at them again. Provided roots eventually form, little time is lost because they continue to grow while attached to their parents.
Layers fail to establish after removal from their parents.	Removed prematurely before the new roots have grown enough to support the tops.	Do not cut the stems before making sure the roots are well developed. Remember that the layers may need attention while establishing themselves in the garden.

Camellias	Euonymus	Ivies	Mulberry	Skimmia
Clematis	Exochorda	Japanese laurel	Myrtle	Smoke bush
Corylopsis	Fatsia	Japanese quince	Passion flower	Vaccinium
Cotoneaster	Firethorn	Jasmine	Peony	Viburnum
Daphne	Heather	Kalmia	Pieris	Virginia creeper
Dogwood	Hibiscus	Kolkwitzia	Privet	Weigela
Elaeagnus	Hollies	Leucothoe	Rose	Wisteria
Enkianthus	Honeysuckle	Lilac	Sea buckthorn	Witch hazel
Escallonia	Hydrangea	Magnolia		

MORE SHRUBS TO GROW FROM LAYERS WHEN A FEW LARGE, WELL-DEVELOPED PLANTS ARE NEEDED TO FILL SPACES

Dogwoods, grown for their bright bark in winter, are not easy to propagate from cuttings, but can be grown from layers.

Dogwoods are often cut back almost to ground level in the spring to encourage fresh growth with bright bark. The long, strong shoots that result make ideal layers, using a technique known as 'stooling'. Peripheral branches are pegged flat along the ground. The young shoots that grow from them are almost buried in a layer of sand and peat when they are about 10cm (4in) high. By the following winter each shoot will have formed roots and can be dug up and separated to form a new plant

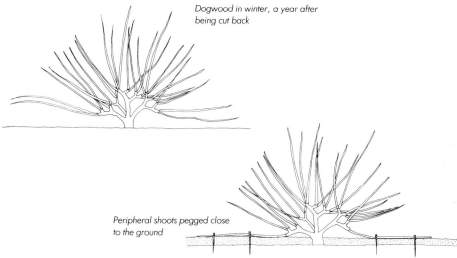

Dogwood in winter, a year after being cut back

Peripheral shoots pegged close to the ground

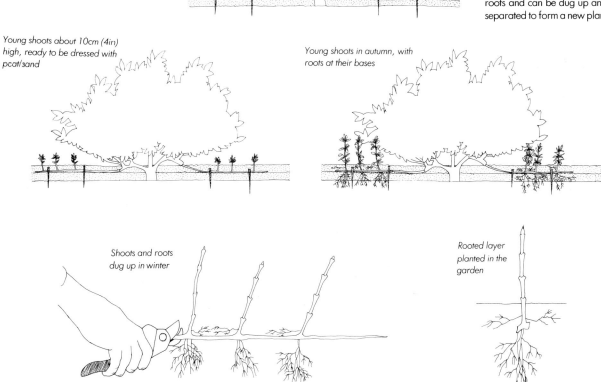

Young shoots about 10cm (4in) high, ready to be dressed with peat/sand

Young shoots in autumn, with roots at their bases

Shoots and roots dug up in winter

Rooted layer planted in the garden

41

MEADOWS AND BAMBOOZALUMS

Ways With Grasses

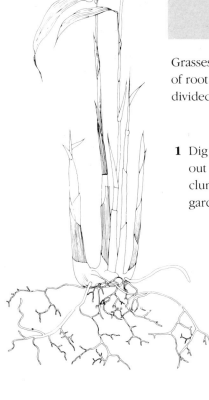

I read somewhere that the famous plantsman E. A. Bowles called his bamboo garden a 'bamboozalum'. Polite gardeners who possessed one during the last few years must have been tempted to call it a 'bamboozal-us', most would have been ruder. The 1980s was the decade of the flowering bamboo – all surrendered to the urge. We once believed that bamboos flowered then died. If only they did! They hang on, looking more derelict and pathetic from one year to the next. But the inclination to flower is passing, and bamboos are again beginning to look beautiful in our gardens.

If you still mistrust them, try a giant perennial grass like *Miscanthus* 'Silver Feather' instead. Then, explore further and discover that, although a few perennial grasses make free with a garden in a style that makes couch grass look reticent, many grow into graceful, well-behaved clumps that never threaten a take-over.

> ### *RECIPE 7*
> ### Springtime Divisions of Grasses and Bamboos

Grasses and bamboos are closely related: one a woody form of the other. Their patterns of root growth and development are similar and, with a rare uniformity, all should be divided when their roots start to grow in the spring.

——— WHAT TO DO ———

1 Dig up grasses in late spring, or fork out bamboo shoots from the edges of clumps. [Plants from a nursery or garden centre could be used. A good

Bamboos and grasses should be divided in the spring when they produce new roots.
This division, made up of two or three old stems with a strong young shoot developing on either side, could be potted up or planted in a nursery bed.

'potful' of a bamboo will often yield a dozen or more divisions.]

2 Divide the grasses into small sheafs of crowns, the bamboos into groups of two or three stems with subterranean buds – you will need to cut them apart with secateurs. The leaves of grasses should be cut off short, the bamboo stems to around 20cm (8in).

3 Using a standard potting compost (see page 18), pot the divisions into the smallest containers able to hold them comfortably.

4 Stand them in a sheltered corner and keep them watered.

5 After three weeks or a month, young roots should begin to encircle the compost. Plant in the garden and mulch with a 4cm (1½in) layer of bark chips.

POSSIBLE PROBLEMS

SYMPTOMS	CAUSE	REMEDY
Plants fail to grow after potting up the divisions.	Insufficient root growth; plants divided too late in the season.	New roots are produced by grasses for a short time each year. Most establish successfully only when divided towards the beginning of this period – in the late spring.

Mr Bowles's Golden Grass from Seed — RECIPE 8

Seeds are introduced in the following chapter and this recipe is jumping the queue a bit. But while looking at grasses – and with Bowles's name as an extra link – this seems an apt moment to point out that many grasses can be grown from seed.

Mr Bowles's Golden Grass will do that by itself. It always self-seeds in a gentle, agreeable way. Once you have it you can increase it by digging up its seedlings, but to make a start beg, borrow or buy seeds and raise a few plants – you'll never regret it.

WHAT TO DO

1 Sow seeds in early summer on the surface of vermiculite (see page 48) in 7cm (2½in) square pots, and plough them lightly beneath the surface.

2 Put the pots in a cold frame and water thoroughly. Replace the framelights, lifting the fronts to provide ventilation on hot days.

3 Prick out the seedlings when they are 2 to 3cm (1in) high. Four in a 9cm (3½in) square pot provides a convenient unit – do as many as you need.

4 Put the pots back in the frame and grow the plants on, keeping the fronts of the frames raised about 30cm (12in) all the time.

5 When the plants are large enough to hold their own, knock them out of the pots, separate one from another and plant them in the garden.

POSSIBLE PROBLEMS

SYMPTOMS	CAUSE	REMEDY
Seeds fail to come up.	a) Poor seed b) Temperature too low.	a) Grass seed is short-lived. Store home-saved seeds as described on page 51. b) Seeds need temperatures above 15°C (59°F) to germinate well. Keep frame closed, except in hot weather, till seedlings appear.

Wood millet is a perennial grass that grows wild in damp, deciduous woodlands. This yellow-leaved variety is equally at home in shade, and few other plants approach its dramatic impact in a darkly shaded corner – and it grows just as well in most situations in a garden. It comes true from seeds, which provide one of the best ways (recipe 8) to start it off once established, and growing happily, it will seed itself. Seedlings will appear in unexpected places, but the plant is always an asset rather than a menace

A QUICK LOOK AT
SOWING SEEDS

Anyone who has ever sprouted beans in a jar would say that growing plants from seeds is child's play. Rinse them with water every morning, keep them warm and, in a few days tip out the sprouts and eat them. Beans are not unusual. Many seeds are just as compliant when kept wet and warm. Others are less easily cajoled, but we'll deal with them later. So, why do people complain that their seeds don't germinate? Or add that when they do, they fall over and die, or grow long and pale and too floppy to cope with.

The reasons lie within the seeds themselves; the causes of failure, in the methods used. Seeds in the soil sense what goes on around them, assess how it affects them, and keep account of changes in their circumstances. They can estimate their depth of burial beneath the surface, perceive the overshadowing presence of other plants, and are aware of the onset of winter and threats of devastating cold. It is the seed's job to wait until its information suggests that an emerging seedling would have a chance of surviving. Then it will germinate.

The gardener's job is to provide conditions that persuade seeds to release seedlings.

Few plants can be brighter and gayer for longer during the summer than annuals. They can be sown outdoors (recipe 10) in the places where they will flower, and some will then establish self-perpetuating colonies that produce annually repeated displays. The picture shows a mixed group of South African annuals including white dimorphothecas, orange ursinias and blue felicias and heliophilas. The Latin names of the two blue-flowered plants refer appropriately to happiness and love of sunshine

THINGS TO THINK ABOUT WHEN SOWING SEEDS

INFLUENCES	NATURAL RESPONSES	HORTICULTURAL SIGNIFICANCE
WATER	Seeds can remain alive in the soil for many years, and are one means by which plants survive hostile circumstances. Only fully hydrated seeds can produce seedlings but other influences such as temperature and exposure to light can override this response. Seeds are particularly sensitive to fluctuations in the water supply.	Dry seeds provide a simple way to send plants from one place to another, and can be stored in packets for years. Many of the widely grown horticultural plants simply need to be sown and watered. It is important to water thoroughly to ensure complete hydration, and to take care not to let the compost become dry at any time between sowing seeds and the appearance of seedlings.
TEMPERATURE	Seeds are able to sense temperatures accurately. They use them as a way of regulating germination – mainly to ensure that seedlings are produced at times of the year which are favourable for their survival. Changes in temperature can also be sensed and used as a measure of the seed's depth of burial beneath the surface of the soil.	Seeds will not germinate when temperatures in the soil are above or below particular levels. The limits vary from one plant to another, but in many long-established garden plants grown from seed, they lead to few practical problems. Less widely grown plants, and more recent introductions, often have more demanding needs, which must be met or the seeds will not germinate.
LIGHT AND DARK	Seeds react to wavelengths in several parts of the visible spectrum. They use changes in light quality as a way of 'seeing' overshadowing vegetation. They measure their depth beneath the soil surface by the intensity of light around them. Some germinate only when exposed to daily illumination. A few germinate only in the dark. Seedlings in low light levels grow excessively long (etiolate) in an attempt to reach brighter conditions.	Although some responses to light are of slight practical importance, it is better not to sow seeds where they will be beneath overhanging foliage. Apart from the very few known to require continuous darkness, seeds should never be covered with opaque, light-excluding materials, or sown so deep that natural light does not reach them. Seedlings sense low light and start to elongate as they emerge. They become long and pale unless grown at high levels (daylight) from a very early stage.

Sowing Seeds in the Ground

When you want to grow lettuces or carrots you sow some seeds in the ground. If they don't come up you can be fairly sure there was something wrong with the seed. Perhaps it was some that had been sitting in a drawer for a year or two; but whatever the reason, the point is that seeds sown outside usually do come up, and with little help from us.

But, apart from vegetables and some hardy annuals, there is a reluctance about sowing like this. We look at the seeds of columbines, canterbury bells, Cupid's dart or coreopsis and think how small and vulnerable they seem. We sow them, and coddle them in a propagator where the artificial warmth brings them up in no time. Then we forget to move them on when we should, and they damp off or grow tall, pale and uninteresting.

Many seeds can look after themselves outdoors better than we can indoors: they are not easily deterred. Delphiniums, 10cm (4in) deep, will force their way to the surface. They may not enjoy it, but nobody would sow them deliberately at such a depth. Corncockle seeds can germinate embedded in ice – it takes them a few weeks but they don't die, they just take longer to appear than they would in kinder conditions.

Three of the recipes (10, 11 and 15) provide details with lists of plants that can be sown in the open ground. Anyone who wants to grow plants with very little effort, and spend almost nothing in the process, should experiment with others.

Many herbaceous perennials can be grown from seed sown out of doors in nursery beds – a simple, labour-saving way to grow these attractive plants.

Choose a sunny position and fork in some peat (or a substitute) and grit. Sow the seeds – just like lettuces or carrots – in shallow drills, marking the ends of the rows with sticks

Sowing in Containers

At home from school I used to make pocket money by helping with the bedding plants on the nursery. These would be sown in seed trays. When the seedlings were just large enough to handle, they would be pricked out – again into seed trays – in batches of sixty or more per tray. By the time we had finished, every bench in several very large greenhouses would be covered with hundreds of boxes of young plants.

Now when I see exactly the same method being taught to amateur gardeners, I wonder why on earth someone who wants a few dozen snapdragons or perhaps a couple of hundred seedlings all told, should be shown how to grow them by the thousand.

CHOOSING THE CONTAINER

All that is needed is something to hold the seedlings until they grow big enough to prick out – a matter of days, sometimes a few weeks. The only containers I ever use now for sowing seeds are small 7cm (2½in) square pots. Fifteen of these pack tightly into a seed tray, which makes a good carrier and stops pots from falling over. One pot holds up to fifty seedling dahlias, a hundred ageratums or two hundred snapdragons. I can use those sort of quantities, not the thousands that seed trays hold.

SOWING THE SEEDS

There is a hallowed way to demonstrate how to sow a packet of seeds – and it uses a seed tray! The seeds are sown on a bed of peat; more peat is sifted over them. They are watered using a careful sprinkling from a fine rose – but, nine times out of ten, the sifted peat and seeds rise in a tide and drift to the sides of the box. The method wastes space, compost, seeds and time and gives good results only with practised skill.

A simplified method is as follows:

A foolproof way to sow seeds – in small square pots in place of seed trays.
The way the seeds are sown depends on their size.

1 **Very small seeds:** Include mulleins, tobacco, begonia. Sprinkle over the surface. They sink into the vermiculite when it is watered.

2 **Medium-sized seeds:** Include pansies, salvias, stocks. Sprinkle over the surface and then 'plough' from side to side with a pointed dibber to bury them lightly.

3 **Large or oddly-shaped seeds:** Include lupins, marigolds, hollyhocks. Spread a thin layer (0.5cm/¼in) of the vermiculite over the compost. Sow the seeds and top up with more vermiculite.

1 Put a handful of potting compost into a 7cm (2½in) square pot: it should be about two-thirds full.

2 Top up with vermiculite. This should be 2.5 to 3cm (1 to 1¼in) deep, reaching to about 0.5cm (¼in) below the rim of the pot.

3 Stick labels into the pots.

4 Sow the seeds in the vermiculite, as shown.

Sow very small seeds on the surface

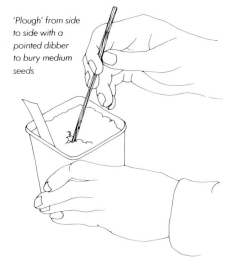

'Plough' from side to side with a pointed dibber to bury medium seeds

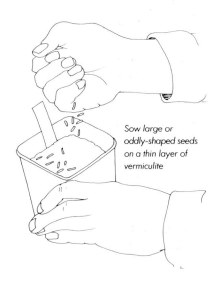

Sow large or oddly-shaped seeds on a thin layer of vermiculite

Cover pots with polystyrene and germinate on a heated bench

48

The pots are packed into seed trays and watered, thoroughly and heavily, using a fine rose on a watering can. They are put wherever they are intended to germinate – a cold frame perhaps, or over soil-warming cables on a greenhouse bench – then covered with a piece of expanded polystyrene. A ceiling tile will do, or one of the trays in which pot plants are delivered to florists.

PRICKING-OUT THE SEEDLINGS

Plants in containers are grown from seed in two stages.

1 Seeds are sown evenly, but relatively thickly, in a small pot, and provided with favourable conditions in which to germinate.

2 Seedlings are pricked out, as soon as they can be handled to give each one the space it needs to develop into a small plant.

The container chosen for the second stage depends on the number of plants needed. Seed trays hold fifty to ninety seedlings. If those are the numbers you need, prick out into seed trays. When you have smaller numbers in mind, use half or quarter trays, or 12cm (4½in) square pots. *Always separate seedlings of different plants or cultivars: never mix them in the same container.*

Dealing with Obstinate Seeds

This chapter started with the suggestion that growing plants from seeds is child's play, quickly followed by the qualification that some are not so easily cajoled. We come to the latter now! Even experienced gardeners are frequently disappointed when trying to persuade these difficult seeds to germinate – the strategies used by plants to control the conditions in which seedlings emerge are so varied. But there is a way round the problem which often works, even for those who know nothing at all about what makes seeds tick.

THE IDIOT'S WAY TO PERSUADE DIFFICULT SEEDS TO GERMINATE

This is a method that I use whenever I expect problems. It amounts to letting nature take its course, while still retaining some control over what is going on.

Ideally, sow these problem seeds immediately after harvesting. When that cannot be done they should be sown as soon as possible. They are not sown in the ordinary way, but mixed with damp vermiculite in a small plastic box with a tight-fitting lid (more details on page 67).

The box is kept where the effects of the changing seasons, and the alternations between day and night, will be experienced by the seeds inside it: for example, in an unheated potting shed, or in a plastic bag or bucket in a corner of the garden. Sooner or later the seeds start to germinate so open the box from time to time and have a look at the seeds. As soon as roots begin to show, spread the contents of the box – vermiculite and seeds together – over the surface of compost, partially filling a small square pot. Put it into a frame or on a bench in a greenhouse, water thoroughly and cover it with a polystyrene tile till the seedlings emerge. After that they can be treated like any other seedlings.

COLLECTING AND STORING SEEDS

A Garden Harvest

Collect seed

Are half the seeds bought ever sown? I wonder how many are left in packets for a year or two, then thrown away? Seeds were once so cheap that the waste hardly mattered. Now the annual seed bill can make quite a hole in your pocket. Seeds left in a packet in a drawer may survive little longer than a year or two. Some die in less than a year – most within five. But, properly stored, they can live on for decades.

How many gardeners collect seeds from their own plants? A great many come true from seed, or at least grow up to play a valued part in any garden. But it's not worth taking trouble to collect seeds from expensive F1 hybrids, unless you can use a ragbag mixture of second-rate versions of their ancestors.

Hang seed in paper bags to dry

RECIPE 9
Collecting Seeds of Delphiniums

Many plants produce masses of seed that will produce seedlings as reliably as anything you can buy. They are a free source of almost unlimited plants.

Sort and clean seeds on folded sheets of white paper

———— WHAT TO DO ————

1 Wait until seeds are mature and about to be shed naturally. The capsules change colour from green to brown or straw when the time is right.
2 Collect the seeds in brown paper bags, preferably on a dry day, by cutting the stems with a pair of secateurs and putting them head first into the bags. Try not to break up the capsules containing the seeds. Label the bags.
3 Tie a string round the open ends of the bags and hang them in a warm, dry, airy place, such as a passageway, a conservatory or a greenhouse.
4 A fortnight later, shake the bags to

dislodge the dry seeds. Avoid breaking up the capsules and other dried bits of plant or mixing seeds with bits of stems, leaves, capsules etc.
5 Empty the bag onto a sheet of white paper. Remove rubbish and deformed seeds, and tip the cleaned seeds into small brown paper envelopes.
6 Write in pencil on the envelopes the names of the plants, dates of collection, and anything else worth noting. Close the envelopes and seal the corners and edges of the flaps with strips of masking tape. Put them in a drawer in a dry room.

Put in small envelopes and label

Storing Seeds from the Garden

Seeds die quickly when they are damp and warm, but stay alive when kept cold and dry (the colder and drier the better). The simplest way to preserve them is in an airtight box containing a mineral dessicant known as silica gel. (Silica gel can be bought from chemists. It is a harmless mineral that can be handled in complete safety.) Keep the box in a refrigerator or, preferably, a freezer.

Store seeds in envelopes in plastic boxes and place in a freezer

——— WHAT TO DO ———

1 Seeds collected from the garden should be kept dry in a warm room for about three months before being stored.
2 Then, find a plastic box with a close-fitting lid, large enough to hold the packets of seeds to be stored. Pour a layer of silica gel 1.5cm (½in) deep into the box. Put the packets of seeds on top and replace the lid.
3 Put the box into the freezer.
4 When you want to sow them, take out as many seeds as you need. Reseal the envelopes with masking tape, and put them back in their boxes in the freezer.
5 The silica gel will eventually become saturated with water. When this happens blue indicator paper included in the pack will turn pink.
6 Take the packets of seed out of the box. Pour the silica gel into a baking dish and dry it in a low oven for three hours. Then put it back in the box; replace the envelopes and return them all to the freezer.

Saturated dessicant can be regenerated by drying in an oven

Storing Left-over Seeds in Packets

——— WHAT TO DO ———

1 Seeds bought in foil packs, or in foil pouches inside a paper packet, should be kept unopened in the freezer till they are wanted.
2 Seeds in paper packets should be stored over silica gel in boxes in the freezer.
3 After sowing, left-over seeds either from foil or paper packets can be stored to use later. Seeds in foil should be transferred to paper envelopes and labelled. Paper packets containing seeds should be sealed with masking tape.
4 Store the packets over silica gel in boxes in the freezer till you need them again.

This simple routine prolongs the life of all small, dry seeds. Most garden plants come into this category, exceptions being the large moist seeds produced by nuts of many kinds, oaks and some maples. The naturally short-lived seeds of primulas, delphiniums, meconopsis etc, left untended in the potting shed, will be dead within a year. Kept dry over silica gel in a cool room, they may survive for three years; in a refrigerator for seven, and in a freezer for twenty or more. Seeds of other, longer-lived plants, kept dry and cold in a freezer, will remain alive for decades or even centuries – heirlooms for your children!

11 TROUBLE-FREE BEDDING PLANTS

Every spring at Oldfield Nurseries, as the curlews started calling, I'd hear customers intoning the gardener's litany ordained for recitation from Lent to May. 'Never again will I buy an annual – they're too much work; they're much too dear. I'll free myself from their dreadful burden and fill my garden with shrubs instead.' A few meant it. Most found the annuals so much brighter, more welcoming and enjoyable than the leafy pleasures of the shrubbery, that they came back for more the following year – still intoning the litany.

But why buy annuals, or any other kind of bedding plant when packets of seed – enough to produce hundreds of plants – can cost less than a loaf of bread? Flip through a seed catalogue and compare the tantalisingly broad range you will find there with the stereotyped selections garden centres have on offer. Enough plants can be raised in a few square yards at the edge of the vegetable patch to fill any garden. Make a simple cold frame and see how the possibilities expand, or commandeer the old greenhouse and make it pay its way at last.

RECIPE 10	**Larkspur Sown in the Borders**

> SOW: Mid-spring
> WEED/THIN: Late spring to mid-summer
> IN FLOWER: Late summer to early autumn

Larkspurs are among the most attractive of all hardy annuals in the garden, and excellent as cut flowers in the house. They can be sown where they will flower and this is an easy way to ring the changes from one year to the next; fill a corner or the spaces between newly planted shrubs. It is worth taking trouble to prepare the ground, and curb any impatience to get started. Seeds sown early when the ground is still cold and wet have less chance than later sowings when warmer, drier ground make a good seed bed easier to prepare.

SOME OTHER HARDY ANNUALS THAT CAN BE SOWN STRAIGHT INTO THEIR FLOWERING POSITIONS

Anchusa*	Corncockle	Gypsophila	Poppy
Californian poppy	Cornflower	Lavatera	Pot marigold
Candytuft	Cosmos	Limnanthes	Scabious*
Chrysanthemum*	Dimorphotheca	Love-in-the-mist	Scarlet flax
Clarkia	Godetia	Mignonette	Summer savory
		Moroccan toadflax	Sunflower
		Night-scented stock	Sweet sultan
*The annual varieties only		Pansy	Virginia stock

WHAT TO DO

1 Choose a space that is open and sunlit.
2 In mid-spring, fork and rake lightly to make a firm seedbed with an even surface.
3 Spread a 1cm (½in) top-dressing of composted bark all over the area to be sown.
4 Draw out shallow (1cm or ½in) drills 15cm (6in) apart with a blunt stick.
5 Sow seeds thinly, a pinch at a time along the drills.
6 Cover by brushing a stick across the crests of the drill and water thoroughly using a fine rose.
7 When seedlings emerge, hand weed, thinning out clumps of seedlings, and filling in gaps; aim for one seedling every 5cm (2in).
8 Three weeks later weed again and thin out the small plants to one every 15cm (6in).

Annuals can be used for bedding, or mixed with perennials or shrubs. Then a small patch – perhaps of a single kind or colour – can be sown to fill a gap or make a planting more colourful for longer. The white poppies have been sown in small groups (recipe 10) amongst flowering shrubs. Their broad white petals are dramatically illuminated by highlights and shadows produced by the foliage of the shrubs

POSSIBLE PROBLEMS		
SYMPTOMS	**CAUSE**	**REMEDY**
Too few plants.	Poor germination; seed spread too thinly.	Use freshly bought or carefully stored seed. Sow thinly but continuously along the drills.
Plants fail to grow properly.	Infertile ground.	Next time scatter a general fertiliser at 100g/sq m (3½oz/sq yd) before applying the top-dressing of mulch/soil conditioner.
Plants grow too vigorously; all leaf, few flowers.	Ground too rich and fertile.	Next time sow the seeds in a less fertile part of the garden.
Weeds take over.	Lack of attention – too much to cope with.	Sow in small, easily reached patches. Make sure the top-dressing is deep enough to keep the weeds down.

RECIPE 11	**Grow Your Own Wallflowers**

SOW: Early summer
WEED/THIN: Mid- to late summer
PLANT: Mid-autumn
IN FLOWER: Spring

OTHER BIENNIALS WHICH CAN BE PRODUCED IN THE SAME WAY

Brompton stock
Canterbury bells
Daisies
Evening primrose
Forget-me-not
Foxglove
Hollyhock
Honesty
Mullein
Opium poppy
Scotch thistle
Siberian wallflower
Sweet William
Teasel

Provided your garden is free from rabbits, which enjoy these plants as a winter-warming relish, wallflowers will add rich colour and fragrance to spring borders. These, and many other hardy biennials, can be grown from seed as easily as cabbages (their close relatives), and you will find a much wider choice of colours and varieties in seed packets than you are likely to get when buying expensive ready-made plants.

WHAT TO DO

1 In early summer find an open sunlit piece of ground – preferably square or rectangular.

2 Apply 100g/sq m (3½oz/sq yd) of a general fertiliser to the ground. Fork it into the top 10cm (4in) and rake lightly to produce an even surface and a good tilth.

3 Draw out drills, 2cm (¾in) deep and 15cm (6in) apart with a blunt stick, and mark the ends with pegs.

4 Sow seeds thinly along the drills, and cover by raking gently across the bed.

5 When the seedlings appear, weed between the rows by hand or with a small hoe.

6 Thin out the young plants when they are 5cm (2in) high, leaving one every 7cm (2½in); use the thinnings to fill in gaps. A square metre of seed bed should hold about 120 plants.

7 During the autumn, drench the plants with a high potash liquid feed (2×). A few days later lift them gently, using a hand fork to loosen the roots, and plant them in the garden.

POSSIBLE PROBLEMS

SYMPTOMS	CAUSE	REMEDY
Poor germination/weak seedlings.	Poor quality seed.	Use only freshly bought or well stored seed.
Plants grow badly, roots deformed.	Club root. Only wallflowers, Siberian wallflowers, Brompton stock and honesty are susceptible to club root.	Sow in clean ground where no brassicas have been grown recently.
Plants grow long and spindly.	Too crowded in the seed bed.	Be careful not to sow too thickly, and thin out the seedlings in good time.

Autumn-sown Snapdragons for Earlier Flowers   *RECIPE 12*

SOW: Early autumn
PRICK OUT: Mid-autumn
PLANT: Mid-spring
IN FLOWER: Late summer to mid-autumn

MORE ANNUALS THAT CAN BE SOWN IN THE AUTUMN AND OVERWINTERED IN A COLD FRAME

Californian poppy
Candytuft
Clarkia
Clary
Corncockle
Godetia
Larkspur
Love-in-the-mist
Mignonette
Pansies
Pot marigold
Shirley poppy
Stock
Sweet pea
Viola

Among the hardy annuals are some so hardy that they can be sown in the autumn in the places where they will flower the following summer. They may not survive very cold winters but, given a reasonable chance, grow into large plants by the spring and come into flower early and abundantly. Pot marigolds, larkspurs, opium poppies and love-in-the-mist are all worth trying in this way and, when they are happy, will usually become colonists, sowing themselves for years to come, wherever they find spaces.

Only a few annuals can be relied on to survive like this. Rain, cold, or both, will kill most others more years than not, especially on heavy, wet soils. Sometimes, too, there are no spaces ready in the autumn where they can be sown.

Snapdragons survive some winters in some gardens but, rather than take a chance I prefer to give them a little protection, and myself a little more flexibility, by sowing seeds under cover in the autumn, and growing them in a cold frame until the worst of the wintry weather is over.

—— WHAT TO DO ——

1 In early autumn set up a cold frame in a sunlit, airy but sheltered corner, covering the lights with bubble polythene.

2 Sow seeds (see page 48) in 7cm (2½in) plastic pots, scattered over the surface of vermiculite and ploughed in.

3 Pack the pots into a seed tray in the frame, water thoroughly with an overhead drench. Cover the pots with expanded polystyrene tiles until the seeds germinate. Ventilate the frame by raising the front 20cm (8in) with a wooden chock. [Alternatively, the seeds could be germinated on a greenhouse bench and moved to the cold frame after pricking out.]

4 When the seedlings are large enough to handle, prick them out into seed trays (sixty seedlings, six rows of ten per tray); return them to the frame.

5 Grow the plants on: keep the front of the frame raised to provide ventilation, closing it only when frost, snow or cold winds are expected.

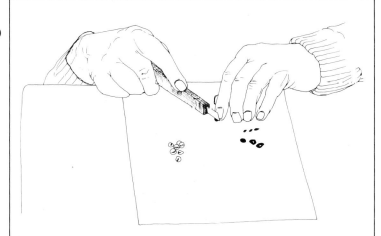

DEALING WITH SWEET PEA AND OTHER SEEDS WHICH HAVE WATERPROOF SEED COATS

Some plants, particularly those in the pea family, produce seeds with waterproof seed coats. These cannot germinate until the seed coat has been damaged, and water can penetrate the seeds and rehydrate the embryos. The sweet pea seeds shown here are being chipped with the point of a knife to remove a small fragment of seed coat.

They will then be left in the open air for about twenty-four hours before being sown in the usual way (see page 48)

Close-up of seed, showing the scar known as the hilum. When chipping seeds, avoid damage to the hilum

6 During very cold weather close the frame and add more insulation, such as an extra layer or two of bubble polythene, old carpets, rush matting.

7 In early spring feed with a heavy drench of high potash liquid feed (1×). Plant out in the garden from mid- to late spring.

POSSIBLE PROBLEMS

SYMPTOMS	CAUSE	REMEDY
Seedlings grow tall and weak with pale leaves.	Etiolated due to low light levels immediately after the seeds germinated.	Remove the polystyrene tile at the *first* signs of germination.
Plants fail to grow large enough.	Sown too late.	Sowing date is very critical: advance it a week or ten days next time.
Plants go mouldy and die during the winter.	Insufficient ventilation, or overwatered.	Keep the front of the frame raised except in severe weather. Water only when sunny, and keep plants a little dry rather than wet at all times.
Plants fail to revive after frost.	Frostbitten	Whenever frosts are severe add more insulation to the frames. Extra layers of bubble polythene or old carpets!

RECIPE 13   Nemesias Sown in a Cool Greenhouse

> SOW/PRICK OUT: Early spring
> TRANSFER TO FRAME: Mid-spring
> STAND OUTSIDE: Late spring
> PLANT: Early summer
> IN FLOWER: Mid-summer to mid-autumn

Panic grips gardeners in late winter and early spring: a feeling that unless they get on with the seed sowing they'll miss the boat. In fact, lethargy – not keenness to make a start – is the key to growing spring-sown annuals. Dark, cold weather during the early months of the year brings problems which disappear as conditions improve.

The annuals that I prefer to sow first are those, like nemesias, which grow naturally where Mediterranean climates prevail: places where cool, damp winters are the growing season. Plants from these places can make the most of chilly, quite inclement conditions – even a touch or two of frost – and grow well in a greenhouse warmed only by soil-heating cables beneath the surface of one or more of the benches.

56

———— WHAT TO DO ————

1 In early spring sow seeds (see page 48) in 7cm (2½in) pots, ploughing them lightly below the surface.

2 Put the pots on a greenhouse bench above soil-heating cables. Water them thoroughly and cover with polystyrene.

3 Check the temperature of the compost around the seeds with a soil thermometer, and try to maintain about 15°C (59°F) during most of the day by adjusting the thermostat controlling the soil-heating cables. Night-time temperatures should not fall below 5°C (41°F). Use a sheet of bubble polythene as a blanket over the seedlings on cold nights.

4 As soon as any sign of germination can be seen, remove the polystyrene tile each day. Replace it each evening until the leaves of the developing seedlings touch its undersurface.

5 Prick out the seedlings into seed trays when their seed leaves are fully expanded, putting sixty seedlings into each tray in six rows of ten. Stand the seed trays on the heated bench.

6 Water regularly but do not overwater.

7 Feed with a high-potash liquid feed (1×) when the leaves of the seedlings in the tray overlap. Continue at fortnightly intervals until the plants go into the garden.

8 In late spring move the trays to a cold frame with the fronts opened each day and closed only during cold nights. Two or three weeks later the trays can be stood in the open and a few days after that the plants will be ready for the garden.

OTHER ANNUALS WHICH CAN BE SOWN EARLY AND GROWN UNDER COOL CONDITIONS

Alyssum
Annual hollyhock
Arctotis
Black-eyed Susan
Californian poppy
Candytuft
China aster
Chinese pink
Chrysanthemum
Clarkia
Clary
Cosmos
Dimorphotheca
Gloriosa daisy
Godetia
Gypsophila
Helichrysum
Larkspur
Lavatera
Mignonette
Moroccan toadflax
Pot marigold
Salpiglossis
Scabious
Snapdragon
Statice
Stock
Summer cypress
Summer savory
Swan river daisy
Sweet sultan
Ten-week stock
Ursinia
Venidium
Viola

POSSIBLE PROBLEMS

SYMPTOMS	CAUSE	REMEDY
Few or no seedlings appear.	Poor germination.	Use only freshly bought or carefully stored seed.
Seedlings collapse and die before pricking out.	Damping off caused by water-borne fungi.	Take care not to overwater; use tap-water rather than water from a butt. Keep the greenhouse as light and well ventilated as possible.
Seedlings wilt and die after pricking out.	Fungal attacks are most probable cause. Death from drought is possible but less likely.	Take care not to overwater, especially during dull weather. However, do not allow peat composts to become so dry that they are difficult to rewet.
Older leaves go tatty, and/or develop tints.	Plants suffering from lack of nutrients.	Increase rate of feeding, doubling the frequency and the concentration for a month.

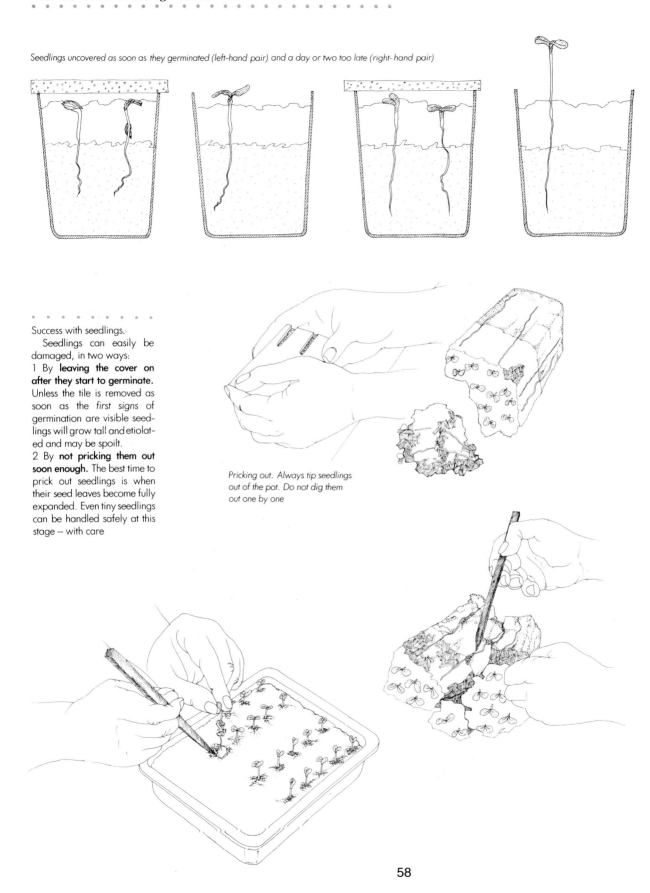

Seedlings uncovered as soon as they germinated (left-hand pair) and a day or two too late (right-hand pair)

Success with seedlings.

Seedlings can easily be damaged, in two ways:

1 By **leaving the cover on after they start to germinate.** Unless the tile is removed as soon as the *first signs* of germination are visible seedlings will grow tall and etiolated and may be spoilt.

2 By **not pricking them out soon enough.** The best time to prick out seedlings is when their seed leaves become fully expanded. Even tiny seedlings can be handled safely at this stage – with care

Pricking out. Always tip seedlings out of the pot. Do not dig them out one by one

58

Petunias Prefer Tropical Conditions

SOW/PRICK OUT: Mid-spring
TRANSFER TO FRAME: Late spring
STAND OUTSIDE: Early summer
PLANT: Mid-summer
IN FLOWER: Late summer to mid-autumn

OTHER ANNUALS
WHICH NEED FROST-
FREE CONDITIONS
AND HIGH
TEMPERATURES TO DO
WELL

Strips and boxes containing lobelias, begonias, tagetes, busy lizzies, petunias and salvias for sale at garden centres are filled with plants that grow naturally in sub-tropical or even tropical climates. They grow successfully only in warm, frost-free conditions, and can be expensive home-grown productions. I like to grow some – especially petunias – because varieties can be bought as seed which are seldom sold as plants, but they have to be carefully fitted in to avoid spending a fortune coddling them. I never sow them early in the year: they will flower from the latter part of the summer right through autumn from a sowing in mid-spring. By then a heated bench in a greenhouse is enough to provide the high temperatures and complete protection from frost which they must have.

African marigold
Ageratum
Begonia
Busy Lizzie
Cock's comb
Coleus
Dahlia
French marigold
Livingstone daisy
Lobelia
Love lies bleeding
Mexican sunflower
Morning glory
Nasturtium
Scarlet salvia
Spider flower
Sunflower
Tobacco plant
Verbena
Zinnia

───── WHAT TO DO ─────

1 Sow seeds (see page 48) in 7cm (2½in) pots in mid-spring. Do not plough in, but water thoroughly to settle the small seeds into the vermiculite after standing the pots on a heated bench in a greenhouse. Cover with polystyrene tiles.

2 Make a canopy of bubble polythene on a light wooden frame above the pots and, using a thermometer suspended inside the canopy, adjust the thermostat controlling the heating cables to maintain air temperatures of about 20°C (68°F).

3 Ventilation of the greenhouse and the canopy will depend on weather conditions outside. Adjust both to keep maximum temperatures below 30°C (86°F).

4 Remove the polystyrene tiles at the first signs of germination. When the seed leaves are fully developed prick the seedlings out into seed trays (six rows of ten plants per tray) and stand the trays on the heated bench. Remove the bubble polythene canopy after a week, replacing it only on cold nights. Aim to maintain minimum temperatures of 15°C (59°F).

5 Open the greenhouse ventilators every day unless the weather is unusually cold, and at night close them, except during warm, humid conditions.

6 When the leaves of adjacent seedlings meet, feed with a high-potash liquid feed (1×), and continue feeding at fortnightly intervals.

7 Move the trays into a cold frame in early summer; in fine weather prop the fronts open during the day, but close at night for the first week. Gradually increase ventilation.

8 Plant out in the garden about three weeks later in early mid-summer.

(*continued*)

59

Many of the most popular annuals, including Busy Lizzies, come from tropical or subtropical parts of the world. Their seedlings depend on high temperatures to germinate and grow well (recipes 14 or 43), and they can turn out to be expensive homemade productions. The secret is to avoid the temptation to sow early when low temperatures and short, dark days create problems for the grower. Plants sown later are much easier, and cheaper, to raise. They soon catch up with the early birds

* * * * * * * * *

POSSIBLE PROBLEMS

SYMPTOMS	CAUSE	REMEDY
Seedlings fail to appear.	Poor quality seed, or temperatures falling too low at night.	Use only recently bought or well-stored seed. Make sure temperatures do not fall below 15°C (59°F).
Seedlings elongate and develop weak stems and pale leaves.	Etiolation due to inadequate light immediately after the seeds germinated.	Be careful to remove the polystyrene tiles as soon as germination starts. Make sure the glass of the greenhouse and the bubble polystyrene canopy are clean.
Seedlings die soon after they germinate.	Fungal attacks causing damping off; or death from over-heating.	Use free-draining compost and take care not to overwater. Keep a check on temperatures: do not let them rise above 30° (86°F).
Younger plants fail to grow and leaves develop tints.	Temperatures too low, or plants suffering from lack of nutrients.	Ensure that night temperatures do not fall lower than 15°C (59°F). Double the frequency and concentration of feeds.

PERENNIALS FROM SEED

12

The Cheapest Way to Fill the Borders

Seed catalogues create dreams of gardens filled with lovely things. They also encourage thoughts that plants whose seeds are sold in colourful paper packets are the only ones that 'ought to be grown from seed'. Annuals and vegetables and a few familiar perennials can safely be grown in this way, but the message is look out for difficulties and disappointments if you stray beyond the limits.

There are tales that 'other plants' need complicated conditions to make them germinate; that they take ages to come into flower; that their seedlings grow up to be only pale shadows of their parents. But numerous perennial plants germinate easily, produce flowers the year after they are sown – and many within six months, and grow up to be every bit as beautiful as their parents. We should not be put off by a difficult, tardy or uncertain minority.

Seeds are a cheap and excellent source of these plants and it is folly not to use them. If you doubt it, start with columbines, polyanthus and delphiniums collected from your garden – or bought, for all these do come in packets. I guarantee the results will be an antidote to any negative thoughts that worthwhile perennials can be found only in pots in garden centres, and you'll soon be looking for opportunities to try less familiar things.

Columbines from a Seedbed in the Garden — *RECIPE 15*

SOW: Late spring to early summer
WEED/THIN: Mid- to late summer
PLANT: Early to late autumn
IN FLOWER: Early summer

Nurseries used to be mostly small, general-purpose affairs. Their produce was sold to people living nearby and they grew a little bit of almost anything their customers might buy – fruit, vegetables, cut flowers and plants of all kinds. When wallflowers and other biennials were sown, a few rows of easy perennials would be put in too, and in the autumn people coming in to stock up on wallflowers, would take away a few delphiniums or columbines or lupins to fill gaps. These would cost more than the wallflowers, but not much more – they had been quick and cheap to grow.

We seem to have forgotten that perennials can be sown in a seed bed out doors. When we want some, we buy them in plastic bags from garden centres, rather than grow a dozen plants from seed. Mindful of advice that we should plant in groups, we buy three of each, but complain about the price as we do so. When we see the threadbare, half hearted display of such a small group, we wish we had bought more.

Many perennials can be sown in a seed bed outdoors, but few more rewardingly than the long-spurred columbines such as the McKana Hybrids. They flower when the spring bulbs are over, but before most herbaceous plants or roses start, and are completely at home in small spaces among other plants.

WHAT TO DO

1 In late spring or early summer make a seed bed in a warm, sunlit corner. Top dress with 2.5cm (1in) of grit (unless the ground is naturally very sandy). Fork it lightly into the top 10cm (4in) of soil, and rake to form an even surface.

2 Make shallow drills 1cm (1½in) deep and 10cm (4in) apart, with a blunt stick and sow the seeds thinly along the drills. Draw the back of a rake across the tops of the drills to bury the seeds.

3 When seedlings are clearly visible weed between the rows with a handfork. Identify the seedlings and thin them out to one every 2 to 3cm (1in). Fill in gaps with surplus seedlings.

4 Repeat this a month later, thinning the seedlings to one every 7 or 8cm (2½ to 3in). A row 1m (3¼ft) long will hold about fifteen plants.

5 Fork out the plants in late autumn or early winter, and plant them where they are to flower.

POSSIBLE PROBLEMS

SYMPTOMS	CAUSE	REMEDY
Very few, or no plants appear.	Germination failure due to poor quality seed, or badly prepared seed bed.	Use only freshly bought seeds, or your own collections that have been carefully stored. Heavy clay soils will need extra grit and/or soil conditioner to form a good tilth.
Seedlings disappear between first and second thinnings.	Eaten by slugs, caterpillars, or some other pest.	Inspect the seed bed one night and destroy slugs or caterpillars. Add a surface layer of grit. Spray bed with liquid slug killer.

OTHER PERENNIALS THAT CAN BE SOWN IN A SEEDBED OUT OF DOORS				
Achillea	Coreopsis	Gay feather	Leopard's bane	Poppy
Arabis	Cupid's dart	Globe thistle	Linaria	Potentilla
Aubrieta	Delphinium	Hollyhock	Lupin	Rose campion
Centaurea	Echinacea	Inula	Mallow	Shasta daisy
Chinese lantern	Gaillardia	Jacob's ladder	Perennial sunflower	Thrift
Coneflower	Geum	Jerusalem cross	Pink	Viola

Blue Poppies from an Early Sowing RECIPE 16

SOW/PRICK OUT: Late winter to early spring
TRANSFER TO FRAME: Mid-spring
POT UP: Early summer
PLANT: Late summer to mid-autumn
IN FLOWER: Early to mid-summer

Using a greenhouse in late winter to protect a few tender pelargoniums, marguerites and fuchsias is an expensive way to keep summer colour going. But a partially filled house can be converted into a usefully occupied house by sowing seeds of hardy perennials.

Seedlings come up rapidly, warmed by soil-heating cables in the bench below them. They can be grown on for a few weeks after pricking out while they get their roots down, and then moved to cooler conditions in a frame when the greenhouse begins to fill up. The perennials have the benefit of a long growing season – many will flower that summer – and all will form large plants by the autumn.

The blue poppies caused a stir when they were discovered growing in the Himalayas, and have stirred gardeners' imaginations ever since. They have a reputation for waywardness and will not thrive in every garden, but if you can find a place for them with plenty of humus in the soil, at least a little shade and an adequate supply of moisture throughout the year, they are well worth trying. Start with *Meconopsis betonicifolia*, and grow it from seed. You have very little to lose that way, and maybe a great deal to gain.

———— WHAT TO DO ————

1 Sow the seeds on vermiculite in 7cm (2½in) pots in late winter and plough them in. Stand them on a greenhouse bench above soil-heating cables. Water and cover with expanded polystyrene tiles.

2 Adjust the thermostat controlling the heating cables so that temperatures in the compost close to the seeds are around 15°C (59°F). Temperatures falling to 5°C (41°F) at night will do more good than harm.

3 Prick out the seedlings when they are large enough to handle. When small numbers are needed put them in square pots, not trays. Five in a 9cm (3½in) pot; nine in a one litre. Put them back on the heated bench.

4 Two or three weeks later, move them to a cold frame. In very mild, sunny weather ventilate from the first day.

Close the frames at night, and during cold spells add extra insulation to prevent temperatures falling below freezing point.

5 In early summer pot the plants individually into 7cm (2½in) square pots and return them to the cold frame. When the tips of the leaves reach the rims of the pots, start feeding every two weeks with a high-potash liquid feed (1×). Keep the fronts of the frames open all the time.

6 Grow the plants in cool, humid conditions within the frame, but do not overwater. In hot, dry weather add an extra sheet of bubble polythene to provide shade, and spray the plants lightly from time to time.

7 Plant out in the autumn and mulch between the plants with a 4cm (1½in) dressing of composted bark.

**OTHER PERENNIALS THAT CAN BE SOWN
IN THE GREENHOUSE IN LATE WINTER**

Achillea	Everlasting pea	Mallow
Aster	Flax	Monkey flower
Astilbe	Gaillardia	Oriental poppy
Balloon flower	Gay Feather	Penstemon
Bear's breeches	Geranium	Pink
Bergamot	Geum	Plume poppy
Cardoon	Globe thistle	Potentilla
Centaurea	Goat's rue	Primula
Chinese lantern	Heuchera	Red hot poker
Coneflower	Hollyhock	Rue
Coreopsis	Iberis	Salvia
Cupid's dart	Incarvillea	Scabious
Dame's violet	Inula	Sea kale
Delphinium	Jacob's ladder	Thalictrum
Diascia	Lewisia	Thrift
Echinacea	Leopard's bane	Toadflax
Erinus	Lobelia	Verbena
Euphorbia	Lupin	Veronica
Evening primrose	Lychnis	Viola

The blue poppy was once an elusively desirable plant, and plant collectors vied with each other to bring seed home. It still has the power to make us gasp at its beauty, but need not be elusive. It sets seed abundantly and, as long as the seed is properly stored (recipe 9), it is not difficult to germinate (recipe 17). The seedlings do need care and the right conditions to grow into flowering plants successfully – but the rewards more than repay those who grow these plants for their trouble

POSSIBLE PROBLEMS

SYMPTOMS	CAUSE	REMEDY
Seedlings fail to appear.	Poor germination due to low seed viability, or inadequate temperatures.	Store seed carefully. Meconopsis seed quickly loses viability under poor conditions. Check temperatures are maintained as suggested.
Seedlings die or disappear after they have germinated.	Destroyed by fungal infections or eaten by slugs.	Encourage healthy growth by maintaining day temperatures around 15°C (59°F). Top-dress pots with grit. Water with tap water.
Plants in frames rot or dwindle away during the summer.	Insufficient ventilation, dry atmosphere, or overwatering.	Keep frames open at all times. Cool, shaded conditions are needed during the summer and especially during heatwaves. Use extra shading and sprays; avoid heavy watering.

Rediscover the Cottage-garden Strains of Polyanthus   *RECIPE 17*

SOW/PRICK OUT: Early to late summer
PLANT: Mid-autumn to early winter
IN FLOWER: Mid-spring

A small quarry overgrown by trees was at one end of my first garden. Guided by memories of the nut plat at Sissinghurst Castle in Kent, I filled parts of it with polyanthus. These had been grown for cut flower in greenhouses on my father's nursery in Bath. They were dug up, most of their leaves were pulled off, and they were crammed into sacks to be carried by British Rail to Dundee in Scotland. 'Old-fashioned' strains, coloured plum and crimson, deep yellow, cream or white, they were split up on arrival and planted out. They did well in spite of this rough treatment and lived beneath my trees for years.

Today's polyanthus have larger flowers: some are a flashy pink, others a wonderfully deep, glowing blue. Their greater size and broader range of colours have been gained at the expense of their constitution. Most are short-lived unless pampered with more tender, loving care than busy gardeners have time to give them. But the older strains of these, and primroses too, are still sold as seed, and if you grow them yourself you will rediscover their robust constitutions.

———— WHAT TO DO ————

1 Sow seeds in mid-summer on the surface of vermiculite in 7cm (3in) pots. Plough lightly and put them in seed trays in a cold frame. Water heavily; cover with expanded polystyrene tiles.

2 When seedlings appear remove the tiles and, when they grow large enough to handle, prick them out into seed trays (sixty seedlings to a tray).

3 Return the trays to the frame. Prop the framelights open at least 30cm (12in) in front, to provide overhead shelter and ventilation. During hot, sunny periods drape an extra sheet of bubble polythene over the frame to keep it cool.

4 Avoid overwatering but never let the trays dry out.

5 In mid-autumn plant the seedlings where they are to flower. Before planting rake in a high-potash general fertiliser at 100g/sq m (3½oz/sq yd).

Achillea	Coneflower	Flax	Inula	Plume poppy	OTHER PERENNIALS WHICH CAN BE GROWN FROM SEED SOWN IN COLD FRAMES IN MID-SUMMER
Agapanthus	Coreopsis	Foxglove	Jacob's ladder	Potentilla	
Alyssum	Cupid's dart	Gaillardia	Leopard's bane	Primula	
Arabis	Dame's violet	Gay feather	Lobelia	Red hot poker	
Aster	Delphinium	Geranium	Lupin	Scabious	
Astilbe	Diascia	Geum	Lychnis	Shasta daisy	
Aubrieta	Echinacea	Globe thistle	Mallow	Thalictrum	
Auricula	Edelweiss	Gypsophila	Meconopsis	Thrift	
Bear's breeches	Erinus	Heuchera	Monkey flower	Toadflax	
Bellflower	Euphorbia	Hollyhock	Mullein	Veronica	
Bergamot	Evening primrose	Hosta	Perennial sunflower	Viola	
Centaurea		Iberis		Woad	
Chinese lantern	Everlasting pea	Iceland poppy	Pink		

Most of these can be grown from seed more easily than polyanthus plants

65

POSSIBLE PROBLEMS		
SYMPTOMS	CAUSE	REMEDY
Seedlings fail to appear.	Poor quality seed. Temperature above level needed for germination.	Polyanthus seed deteriorates rapidly under poor conditions: it must be well stored. Cool or mild conditions are needed for good germination. Use shading when necessary.
Seedlings disappear after germination.	Destroyed by slugs, or damped off due to fungal infection.	Check for slugs and spray surface beneath the framelight with liquid slug killer. Avoid overwatering. Use tap water until the seedlings are well established.
Seedlings fail to grow well, develop very unevenly.	Compost not sufficiently well watered to remain consistently moist. Atmosphere too dry.	The right balance between over- and under-watering is vitally important for success. During hot, sunny spells spray the plants and the ground around them once or twice a day.

RECIPE 18 — Christmas Roses Sown in a Plastic Box

SOW: Mid-summer

PRICK OUT: Early- to late winter

POT UP: Late winter to early spring

PLANT: Mid- to late summer

IN FLOWER: Late winter

Most of the seedlings will take another year before they start to flower

Until now, recipes for sowing seeds have followed the formula 'add water, keep warm and watch them grow'. But the seeds of some plants have more complex needs.

Dedicated gardeners spend much of their lives unravelling the processes which control seed germination; devising one treatment for some plants, another for others. Beginners with mortgages, babies and golf handicaps foremost in their thoughts have other things to think about. They could be forgiven for avoiding involvement with plants whose seeds are so demanding.

That would be a pity, because it would mean missing the pleasure of growing plants as beautiful as the Christmas roses, Peruvian lilies and alpine gentians. It would also be a pity because a simple and elegantly satisfying way to deal with the problem exists. It may not always work, but then nothing to do with gardening *always* works – not least the devices that experts depend on to persuade awkward seeds to germinate! No special skills or experience are needed to give this method a go.

────────── WHAT TO DO ──────────

1 Collect seeds as soon as they ripen: hellebore seeds are shed just *before* the capsules turn straw-coloured. So watch them daily!

2 Put a handful of vermiculite in a nylon kitchen sieve and dip it briefly into a bowl of water. Remove it and let surplus water drain away.

3 Partially fill a small plastic container with the vermiculite, add the seeds, and stir well to mix. Close the lid.

4 Place the container in a cold frame, or put it in a plastic bucket covered with bubble polythene. Stand it in a sheltered corner.

5 Open the box every two or three weeks from late autumn onwards and examine the seeds.

6 They will start to germinate sometime during the winter. As soon as they do so tip them with the vermiculite onto a layer of potting compost in a small pot.

Put the pot in a cold frame, and water heavily.

7 Prick out the seedlings when the seed leaves are fully expanded. Use trays or pots depending on the number of plants you want. Return them to the cold frame and keep the front permanently raised for ventilation.

8 In the spring, pot the seedlings individually into 7cm (2½in) pots and return them to the frame. Grow them on with the fronts raised for ventilation and drape an extra sheet of bubble polythene over the frame to keep it cool during hot sunny weather.

9 Plant in late summer after preparing the ground by forking a dressing of composted bark into the top 10cm (4in). Mulch with a 3cm (1¼in) layer of chipped bark; protect/mark the plants with twigs and keep them weedfree and well watered till autumn.

> **SEEDS OF PERENNIALS WITH ERRATIC OR DELAYED GERMINATION, WHICH RESPOND TO THE PLASTIC BOX TREATMENT**
>
> Aethionema
> Anemone
> Bear's breeches
> Bugbane
> Clematis
> Cyclamen
> Eryngium
> Euphorbia
> Gentian
> Geranium
> Globe flower
> Iris
> Lady's mantle
> Lewisia
> Masterwort
> Monkshood
> Pasque flower
> Peruvian lily
> Peonies

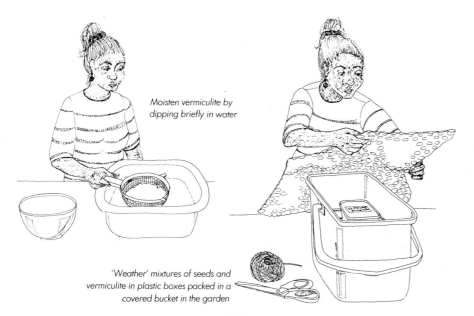

Moisten vermiculite by dipping briefly in water

'Weather' mixtures of seeds and vermiculite in plastic boxes packed in a covered bucket in the garden

Succeeding with difficult seeds. Seedlings of many plants — amongst them hellebores — appear only after the seed has been lying in the soil for months or even years. Gardeners have learnt that these seeds are quite likely to be eaten by predators, or lost in other ways, before they germinate, unless, something is done to prevent that happening

67

POSSIBLE PROBLEMS		
SYMPTOMS	**CAUSE**	**REMEDY**
Vermiculite dries up before the seeds germinate.	Boxes insufficiently airtight.	Re-moisten vermiculite. Next time use a more airtight container.

RECIPE 19 — Starting a Bluebell Wood

SOW: Mid-summer
PRICK OUT: Early spring
STAND OUTSIDE: Late spring
TRANSFER TO FRAME: Early autumn
STAND OUTSIDE: Late spring
PLANT: Early to mid-autumn following year
IN FLOWER: Late spring

Seedlings pricked out in late winter

The plant during its second winter/spring

Sowing bluebells – and other bulbs.

The seed of many bulbous plants should be sown as soon after it ripens as possible. Seedlings appear in late autumn, or winter, and take several years to grow large enough to produce flowers

For years I never bothered to sow seeds produced by bulbs which had flowered in my garden, convinced they would take so long to produce flowers that my patience would wither on the way. I broke my fast with seeds collected from some gladioli I had dug up in order to store the corms. These germinated profusely a few weeks after sowing them in a frame the following spring, grew throughout the summer and, after a winter's rest, flowered barely eighteen months after they were sown. Gladioli are the offspring of several different species, and these seedlings were a revelation of the colours and forms that exist behind the well-regulated facades of 'approved' garden varieties.

Seedlings from other bulbs are less likely to cause so much surprise, but the effort of growing them is rewarded by the large number of bulbs that can be produced. Scattering bluebell seeds on the ground beneath some trees is no way to start a bluebell wood, but collecting seeds from a group of plants, sowing them and taking care of them till they form bulbs is a very effective way to go about it.

——— WHAT TO DO ———

1 Collect fully ripe seeds in mid-summer and hang the gaping capsules upside-down in a paper bag.

2 Sow within a week on vermiculite in 7cm (2½in) pots (page 48), and plough them in; stand them in a frame and water. [These seeds could be sown, like the Christmas roses, in a plastic box. This is an alternative, rather more chancy, way to deal with dilatory seeds.] Continue to water them occasionally throughout the summer.

3 The seeds will germinate in mid-winter. Prick them out in early spring when they are about 2.5cm (1in) high. Twenty-five seedlings will fit into a one litre pot.

4 Stand the pots on a level surface outside and water them when necessary till the leaves die down in early summer.

5 In mid-autumn go through the pots, remove any weeds and apply a sprinkling of high-potash fertiliser. Put the pots back in the cold frame and

leave them there through the winter.

6 Stand them outside again in the spring. When the leaves die in the autumn turn the bulbs out of the pots, and plant them where you want them to flower.

Agapanthus*
Camassia
Crocus
Dierama
Eremurus
Erythronium
Fritillary
Gladiolus*
Grape hyacinth
Ixia*
Lily (*a*)
Lily of the valley
Nomocharis
Scilla
Sisyrinchium
Snowdrop
Summer hyacinth*
Tulip

*Keep the seed till early spring before sowing it. (*a*) Sow seeds of *L. martagon* in the autumn, but no seedlings will appear before the spring or early summer. Do not sow seeds of *L. regale* until early spring.

POSSIBLE PROBLEMS

SYMPTOMS	CAUSE	REMEDY
Seeds fail to germinate during the winter.	Germination is strongly affected by variations in temperature, and will not always occur as and when expected.	Leave the seeds in their pots in the cold frame, and wait for seedlings to appear.
Seedlings fail to reappear after the first year.	Died during the summer resting period.	Seedlings should be neither bone dry, nor too wet during summer. Shelter them under cover of a framelight during wet spells. Water occasionally during heatwaves.

GROWING LILIES FROM SEED

Some lilies produce seed that germinates within a few weeks of being sown, and these need no special treatment. Examples include *Lilium amabile, bulbiferum, davidii* and many hybrids with turkscap and upward-facing flowers, *formosanum, longifolium, pumilum, regale* and most of the hybrid trumpet lilies, and *tigrinum*.

Other lilies germinate in two stages:

1 The roots appear soon after they have been sown, provided they are kept warm.

2 The leaves will not start to grow until the seeds have been through a period of low temperatures, followed by warmer conditions (as spring follows winter naturally). Lilies which follow this pattern of germination include: *Lilium auratum, candidum* (Madonna lily), *hansonii, henryii, japonicum, martagon* and some of the smaller-flowered turkscap hybrids, *pyrenaicum* and *szovitsianum*.

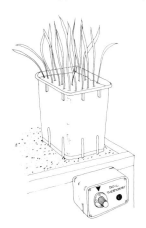

Seedlings on a heated bench after completion of processes of germination

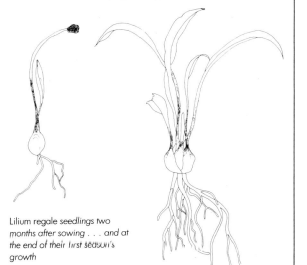

Lilium regale seedlings two months after sowing . . . and at the end of their first season's growth

13

SHRUBS FROM SEED: UNEXPECTEDLY EASY

Pheasants were thin on the ground in the countryside around my garden in Scotland. Hearing that they were partial to Chinese nutmeg berries, I had thought that a few of the shrubs planted in the garden might tempt a bird or two to come in. Improved prospects of roast pheasant for dinner were not far from my mind. I sowed seeds collected from a friend's garden, not expecting much to happen for rather a long time. Within a week or two they germinated in hundreds. I pricked out some seedlings, potted up young plants when they were ready, and several dozen were large enough to plant by the autumn. The pheasants never showed any interest in them – or their berries!

Growing shrubs from seed can be a test of patience and skill, but frequently neither is needed. Even so, the option is seldom considered, perhaps because of a belief that they will take ages to mature and produce flowers; perhaps because space is limited, and we think about seeds mainly when we are looking for many rather than a few plants. But many attractive shrubs grow large enough to produce flowers after only two or three years, and there are situations – plans to plant a hedge are an example – when lots of plants are needed.

RECIPE 20	Sowing a Hedge of 'Munstead' Lavender

Traditionally lavenders have been grown from cuttings taken in early autumn – an uncertain process that works wonderfully well one year, and fails totally another. The first shoots to appear in the spring can also be used to make cuttings (see page 82), but there is yet another way.

Lavender flowers are followed by lavender seed. Hidden within the spikes of dead heads, it lies unnoticed, but take a dry spike, rub it vigorously between your palms and dark, shining seeds will be seen amongst the debris. These grow up to be just like their parents: seeds taken from 'Munstead' produce short, compact bushes with mauve flowers, those from Old English lavender grow into tall, late-flowering bushes, with silvery, fragrant foliage and long spikes of misty lavender flowers.

The seeds germinate readily enough but seedlings emerge in a ragged, disorderly way. It can take several weeks to reach a quorum, and even then half or more will still be waiting to make their entrance. This should not be a problem but you will need to sow rather generously.

———— WHAT TO DO ————

1 Sow seeds in 7cm (2½in) pots in late spring, and put the containers in a seed tray in a cold frame.

2 Water thoroughly and cover with an expanded polystyrene tile. Close the frame on cold nights but keep the

70

front raised at other times.

3 Remove the tile when the first seedlings appear, but leave them undisturbed until enough have emerged to meet your needs.

4 Tip the seedlings out of the pot. Separate one from another without damaging the roots. These will be long and well developed even on small seedlings.

5 Pot the seedlings separately in 7cm

2½in) pots and put them back in the frame. Water them, replace the framelight, and raise the front around 30cm (12in).

6 Six weeks later move the seedlings out of the frame, and start feeding at fortnightly intervals with a high-potash liquid feed (1×). They can be planted in mid-autumn, or returned to the shelter of the frame for the winter and planted out early the following spring.

> **MANY OTHER SHRUBS CAN BE GROWN FROM SEED**
>
> Uneven germination over a long period is typical of lavender, but most of the following produce seedlings rapidly and more evenly:
>
> Abutilon
> Broom*
> Cistus*
> Eccremocarpus
> Exochorda
> Fatsia
> Fuchsia
> Genista*
> Gorse*
> Hydrangea
> Chinese nutmeg
> Tree lupin*
> Pittosporum
> Sage*
> Skimmia
> Tree mallow
>
> *Seeds produced by these shrubs very often have hard seed coats. Chip off a small part of the 'shell' before sowing them (see page 55).

POSSIBLE PROBLEMS

SYMPTOMS	CAUSE	REMEDY
Seedlings disappear after germinating.	Damping off due to attack by soil-/water-borne fungi	Take great care not to overwater. Ventilate the frame freely whenever possible. Water with tap water; use an open, free-draining potting compost.
Seedlings in pots develop purple tints or grow very slowly.	Starvation due to lack of nutrients. Particularly likely after very wet weather.	Double the frequency and the concentration of liquid feeds.

Rhododendrons from Seed   *RECIPE 21*

Rhododendrons are such over-enthusiastic seed producers that the following year's display of flowers may be reduced if the seed capsules are not laboriously removed. A few can be left until they ripen, then collected and sown. These are plants that hybridise with abandon! Their seedlings may look little like their parent, but many, especially among the smaller species and hybrids, will be valuable garden plants.

———— WHAT TO DO ————

1 Collect seed during the winter, preferably as the capsules open, but seeds from green capsules usually germinate.

2 Partially fill 7cm (2½in) square pots with ericaceous potting compost and cover with a 1cm (½in) layer of 50/50 moss peat and perlite. Water well.

3 Sprinkle the seeds over the surface and put the pots in an electrically heated propagator in a greenhouse or conservatory.

4 Drape a sheet of bubble polythene over the top of the propagator. Adjust the thermostat to maintain temperatures around 20°C (68°F).

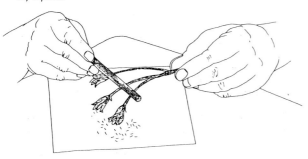

Ripe seed can be tapped out of
dry capsules

Sow seed on the surface
in a small square pot

Growing rhododendrons from
seed.

Rhododendrons produce
tiny seeds in capsules. These
usually mature during late
winter or early spring, splitting
open as they dry out to release
the seeds.

Even seeds extracted from
green capsules will germinate
provided they are sown at
once

. . . .

5 Prick out the seedlings when they are
1cm (½in) high, putting five into a
9cm (3½in) square pot. Keep them
growing in an unheated, fully ventilated
greenhouse all summer.

6 In early autumn move the pots of
seedlings into a north-facing cold
frame. Water sparingly through the
winter and ventilate whenever it is not
freezing hard.

7 Pot the seedlings individually into 7cm
(2½in) pots the following spring.
Return them to the frame and plant in
a nursery bed during the summer.

OTHER SHRUBS WHICH
CAN BE TREATED IN A
SIMILAR WAY

Cassiope
Enkianthus
Heather
Strawberry tree

POSSIBLE PROBLEMS

SYMPTOMS	CAUSE	REMEDY
Seedlings collapse and die soon after they germinate.	Too damp, or too hot and humid.	The watering/ventilation balance is very critical. Water sparingly; provide plenty of ventilation, but do not allow to dry out.
Plants die during the first winter.	Plants either too small, or kept too damp, or both.	Young plants should be kept growing as long as possible during the summer. Be careful not to overwater them in the frame.

RECIPE 22   *Daphne mezereum* **from Seed**

Most of the mezereons on sale at garden centres are large, unmistakably long-limbed
bushes, which have been grown rapidly on fertile soils with a high water table. Moved to
the less benevolent conditions of the average garden, they suffer deprivation and
struggle to establish.

The beautiful fragrant flowers in late winter, which are the main reason for growing
these plants, are followed by translucent berries – poisonous to us, but relished by
birds. Gather these before the birds find them and grow your own mezereons. Tougher
and more adaptable than the long-limbed garden centre beauties, they will grown
better under rougher conditions. Planted in dark corners or beneath the shade of trees,
they will flower before most other shrubs have started to grow.

WHAT TO DO

1 Keep a close watch on the shrubs as the berries ripen, and, just as they start to turn colour – usually in late summer – pick them off. (Pink-flowered bushes produce red berries; white produce yellow. Both come true from seed.)

2 Each berry contains one large seed. Remove these by hand. Remember that the plants are poisonous, so don't lick your fingers! Wear rubber gloves if that makes you feel safer.

3 Sow the seeds at once in one litre plastic pots. Partially fill the pot with a general-purpose potting compost; add a layer, 1cm (½in) deep, of 50/50 grit/vermiculite; scatter the seeds over the

Raising daphnes from seed.

These shrubs produce berries which the birds eat and then distribute the seeds. Berries should be collected as they ripen – but before the birds find them – and the seeds extracted and sown at once. Seedlings seldom appear until the following spring or even later, and should then be potted up individually while still small. Later, they can be planted out in a nursery to grow on for a year – two if necessary

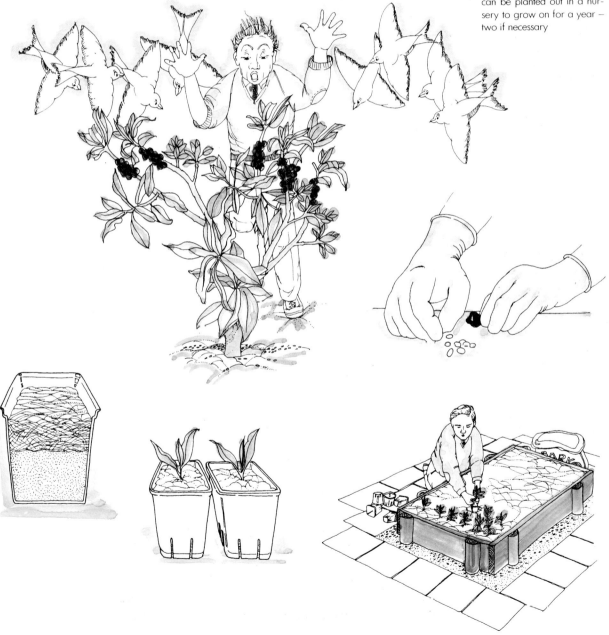

73

surface and top up with 2cm (¾in) of the same mixture to fill the pot to within 1cm (½in) of its rim.

4 Stand the pot outdoors in a sheltered place, and avoid knocking over. Water occasionally during dry weather.

5 The seeds will start to germinated the following spring. When they first appear, put the pot in a ventilated cold frame, and leave until most of the seedlings are about 3cm (1¼in) high.

6 Tip out the seedlings; separate them carefully and pot them individually in 7cm (2½in) pots. Return to frame.

7 Eight to ten weeks later prepare a nursery bed in a fertile, sheltered corner of the garden. Top-dress with 5cm (2in) of mulch/soil conditioner and fork it lightly into the top 10cm (4in) of soil.

8 Plant the seedlings 15cm (6in) apart and mulch them with shredded bark 3cm (1¼in) deep. Keep weeded.

9 The following spring apply a general fertiliser between the rows at 100g/sq m (3½oz/sq yd). Renew the bark mulch. Water when necessary during the summer and plant the young plants out early in the winter.

POSSIBLE PROBLEMS

SYMPTOMS	CAUSE	REMEDY
Berries disappear before they are collected.	Eaten by birds – most probably by greenfinches.	Watch the bushes carefully for signs of depletion and collect as soon as the number of berries starts to fall.
Seedlings fail to appear in the first spring.	The seeds need alternating mild and cold temperatures to germinate. Some years do not provide these changes at the right times or in effective sequences.	Leave the seeds in the pots until the next spring; most will produce seedlings then.

OTHER SHRUBS WHICH PRODUCE SEEDS

Many other shrubs produce seeds, often in berries*, which should be sown as soon as they are gathered and left outside in pots exposed to the elements. Some will germinate the first spring after sowing; others are more likely to appear a year later. These can also be sown in plastic boxes (recipe 18).

Berberis*	Elder*	Laurel	Smoke bush
Caryopteris	Firethorn*	Pernettya*	Tree peony
Clematis	Honeysuckle*	Rose*	Vaccinium*
Cotoneaster*	Japanese quince	Skimmia*	Viburnum*

TREES FROM SEED

14

'Tall Oaks from Little Acorns Grow'

Trees can be trouble: they grow large and intrude on neighbours' space and light. An incident I remember followed the gift of a horse-chestnut tree to the research station where I worked in Scotland. It was a fine sapling, ceremoniously presented by a senior member of staff. His daughter had grown it from a conker: sentiment made a home execution impossible. It had no place on the research station either, and members of staff played pass the parcel as they contrived to have it moved from their experimental plots to another's. The game ended when a new, more ruthless director ordered its destruction.

Half a dozen, a dozen – twenty or more seedling trees need little space. They grow and their demands grow with them. We shrink from throwing any away, neighbours and friends soon tire of them as gifts, and they become neglected. They stop growing, until one day we plant them in a hurry, one almost on top of another in an unsuitable space. A wiser decision might be to use them to fuel a bonfire.

Acorn to Oak Tree	*RECIPE 23*

Before embarking on trees, think before you sow. Decide how many you want and resist the temptation to grow more. Then treat those you do grow like fighting cocks. Trees, more than most plants, can hang on in adversity, but they revel in the good times. Young oaks left all summer in the seed bed may scarcely grow to 15cm (6in). Potted up, grown on in a greenhouse, and fed and watered, they will reach 150cm (5ft) – ready, at a year old to be planted where they are to grow.

———— WHAT TO DO ————

1 Immediately after collection put acorns into a transparent, heavy-gauge polythene bag, half-filled with barely moist peat or shredded bark.

2 Put the bag in a cool place, unheated but more or less frostproof, such as a garage, cellar or shed.

3 Check them once a month. When they have roots 5cm (2in) long, select the most vigorous and pot 30 per cent more than you will need into 7cm (2in) pots with an all-purpose potting

compost. Stand them on an unheated bench in a well-ventilated greenhouse. [Seedlings are grown in a greenhouse in this recipe. They could be grown in pots standing in the open, but would grow much more slowly and are more likely to be eaten by mice or birds.]

4 When the seedlings reach 15cm (6in) move the best of them into 12cm (4½in) pots and put them back on the greenhouse bench. Throw away the rest.

Many trees produce seeds in the autumn that germinate the following spring. Different kinds need to be treated in different ways.

Group 1 Large moist seeds, eg acorns, chestnuts, walnuts etc, can be stored through the winter in plastic bags in barely damp peat in a garage or cellar (cold but almost frost-proof).

Group 2 Small, dry seeds, eg birch, hornbeam, alder should be mixed with moist vermiculite in small plastic boxes packed in a bucket covered with polythene out of doors.

Group 3 Seeds in berries, eg rowan, holly, thorns should be extracted and mixed with vermiculite in small plastic boxes. These should be put in an airing cupboard for six weeks and them moved to a plastic bucket out of doors.

Seedling trees can be planted in a nursery, but grow much faster if potted up and looked after well in a greenhouse during their first summer.

5 In mid-summer pick out the number you need. Pot them into five-litre pots and stand them on the floor of the greenhouse.

6 Start feeding one month later with a high-potash liquid feed (2×), repeat every three weeks until the leaves start to fall in late autumn.

7 As the leaves start to drop, remove the pots from the greenhouse and stand them outside. The young trees can be planted during the winter.

POSSIBLE PROBLEMS

SYMPTOMS	CAUSE	REMEDY
Acorns go rotten in the bags.	Bark or peat too wet. Killed by low temperatures.	Storage material must not be bone dry, but should barely feel damp. Temperature should not fall below 0°C (32°F).
Seedlings fail to appear after they are potted up.	Eaten by field mice, birds or squirrels.	Use mouse traps with 'natural' baits, eg sunflower seeds rather than cheese. And cover with wire netting.
Plants do not grow as well as expected.	Most likely causes are shortage of water, or lack of nutrients.	Plants must be well watered throughout the summer. Increase the rate of feeding if the lower leaves drop or change colour.

Many have small seeds, less easy to handle individually than acorns. When these start to germinate in spring, the mixture of peat and seeds in the bags should be spread over the surface of compost in pots. The seedlings can be potted up individually later. Some, marked (*a*), may not germinate until the second spring after collection.

Ash	False acacia	Horse-chestnut	Maple	Rowan(*a*)	Thorn(*a*)
Beech	Hazel/cob nut	Larch	Mulberry	Sweet chestnut	Thuya
Birch	Holly(*a*)	Lawson's cypress	Pine	Spruce	Yew(*a*)
Cherry(*a*)	Hornbeam	Magnolia			

INTRODUCING CUTTINGS

15

Responses to propagation courses at the Garden School reveal a widespread belief that *real* gardeners do it with cuttings. Division seems too easy; seeds are a bit 'iffy' and might take too long; but cuttings, participants declare, are what they've come to learn about. I would never agree with that, but, as an encouragement to the audience to hang on till the end, cuttings are not usually covered on our courses until lunch is over. In case you should feel the same way, recipes for home-made cuttings appear after those for divisions and seeds have come and gone.

Cuttings convey a certain aura of mystery. Why should a snippet of a shoot, which would never normally produce roots, obediently do so just because that's what is expected of it? It is almost as though the gardener acquires power over the plant to make it do as he or she wishes. Nothing could be further from the truth.

Plants organise themselves according to what is going on around them, and their needs for survival. A shoot will produce leaves, stems, buds – perhaps flowers and seeds – in other words, do all the things a shoot might be expected to do, so long as it is left undisturbed. Cut off from the rest of the plant it responds to the change in its situation by forming roots – because it has no hope of surviving unless it does so. The gardener may help a little by applying a rooting hormone, which reinforces the natural processes which lead to root production. But it is far more important to provide the conditions which will let the cuttings remain alive and functioning. When that is done successfully, they will do the rest themselves.

What Makes a Cutting?

So many different bits of plants are used as cuttings that it might seem any part would do. That is almost true, with one qualification. The cutting must contain cells capable of dividing, for cell division is the start of all the processes by which plants grow and develop.

Young stems and roots are almost always a good bet, especially the places where leaves and stems are joined – known as the nodes. Flowers, the fleshy parts of fruits, tubers and most leaves seldom contain cells that are able to divide, and are therefore rarely used as cuttings.

Most plants can be propagated from cuttings in various ways and the one chosen depends on opportunity, the equipment and facilities available, and the amount of time that can be spent looking after the cuttings – in other words, your convenience.

As with seeds, the methods that make fewest demands are those that use little or no equipment. The most demanding are those that provide for the cutting's every need by minutely controlling its surroundings. Skilfully used, the latter extend the range of possible plants, increase the proportions that form roots, and speed up events. How-

TYPES OF CUTTINGS

TYPE OF CUTTING	ADVANTAGES	DISADVANTAGES
Hardwood Deciduous (late autumn)	These need minimum care and no equipment. Cuttings in a sunny, well-drained bed in the open ground produce roots during the winter, and develop shoots during summer.	Cuttings may not be successful in very wet or cold winters. Only a limited range of hardy shrubs can be propagated in this way.
Hardwood Evergreen inc. conifers (late autumn)	Cuttings in a cold frame will form roots during the winter. They will need to be watered from time to time, but require little other attention.	In cold winters losses may be high when propagating many of the more attractive, less reliably hardy plants. Cuttings in early spring (see below) may do better.
Basal cuttings of herbaceous plants (mid- to late spring)	Produce roots rapidly and reliably in a cold frame. Need regular attention with regard to watering/shading/ventilation, but this is not demanding and is needed for only a short time.	Very few problems or disadvantages. Once roots have been formed the cuttings need to be potted up quickly and grown on in good condition to avoid setbacks.
Mature cuttings of semi-tender and silver-leaved shrubs (early autumn)	This is a good way to insure the survival of less hardy plants. Cuttings can be set up on a greenhouse bench and overwintered in a cold greenhouse. Roots are readily produced and management is not demanding.	Losses will occur during cold winters and among more tender plants. Success rates are greatly improved when cuttings can be overwintered on a gently warmed bench – just sufficient to keep them frost-free.
Semi-mature (or summer) cuttings of shrubs, herbs, alpines etc (mid- to late summer)	Abundant cuttings are readily available. Can be set up in cold frames, using the natural warmth of summer. Cuttings of many different plants root well at this time of year and make few demands on the skills of the propagator.	Cuttings in frames exposed to sunshine will need daily attention. Less demanding systems can be used, for example a heavily shaded frame, but cuttings will not produce roots so rapidly in these.
Tip cuttings (late spring)	Can be a very rapid way to reproduce young shrubs. A few kinds can be propagated in this way only.	The immature young shoots are vulnerable to heat and drought. Skilful attention is needed to maintain favourable conditions.
Root cuttings (late winter)	Some plants can be propagated in this way which are slow or difficult by other methods. A cold frame can be used, but a heated bench in a greenhouse is likely to produce better results.	Success rates can be variable from one year to another for no obvious reason. Newly formed plants are small and may die if not carefully handled.
Single bud cuttings (at various times of the year)	Makes very economical use of material. Can be useful when only a few shoots are available. One of the simplest and best ways to propagate lilies.	Shoots of shrubs and perennial plants from single buds can be very small and may require special care during their first winter – eg a place on a greenhouse bench above heating cables, not a cold frame.
Hardwood Evergreen (early spring)	Produce roots rapidly and can be potted up and grown on quickly to make good plants by the autumn.	Cuttings of many shrubs at this time do well only under mist propagation and with bottom heat (from soil-heating cables).

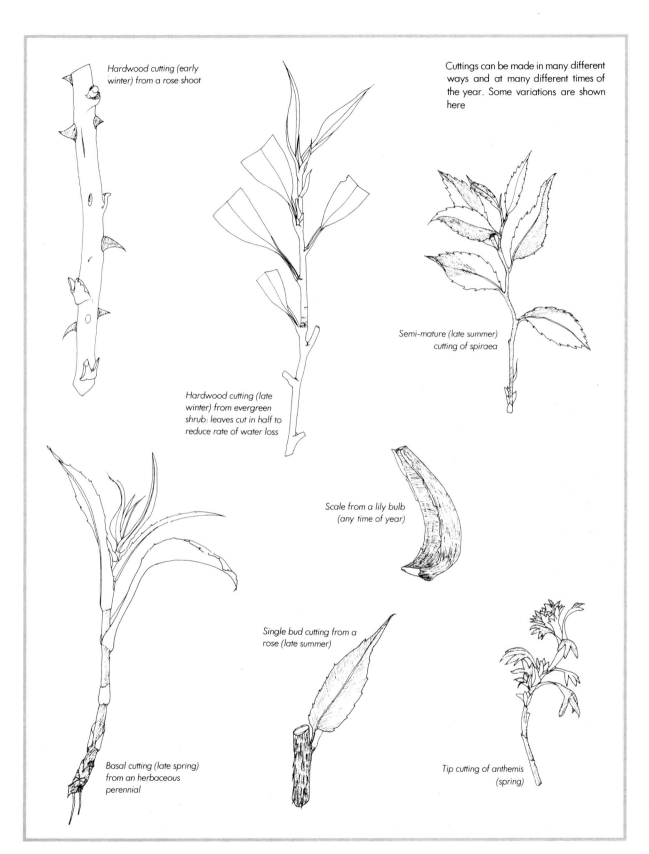

Hardwood cutting (early winter) from a rose shoot

Cuttings can be made in many different ways and at many different times of the year. Some variations are shown here

Semi-mature (late summer) cutting of spiraea

Hardwood cutting (late winter) from evergreen shrub: leaves cut in half to reduce rate of water loss

Scale from a lily bulb (any time of year)

Basal cutting (late spring) from an herbaceous perennial

Single bud cutting from a rose (late summer)

Tip cutting of anthemis (spring)

ever, unskilful or inattentive management will produce far worse results than more modest efforts using simpler methods. In the table on page 78 simpler methods are entered first; more complex ones later.

Contributors to Success

Water is the first need. Without it the cutting wilts, and when it wilts it stops functioning and, if it wilts severely, it dies. The cut end of the stem can take up water, less efficiently than the missing roots but enough, with some help, to satisfy basic needs. The leaves can also absorb water, but in dry air they are the major cause of loss. The most effective way to balance water conservation in leafy cuttings is to keep them in an atmosphere saturated with water vapour. That is why cuttings are enclosed in damp conditions and sprayed with water from time to time, with their stems stuck into wet cutting compost.

But staying alive is not enough: cuttings must be able to function and grow, and for that they need energy obtained from sunlight. It has to be bright light, not the subdued levels of a sick room.

To summarise then, success depends on:

a) *providing as much light as possible;* and

b) *keeping the cuttings enclosed in a saturated atmosphere.*

The inevitable result of doing both without restraint is a 'Turkish bath' that steams the life out of the cuttings. So how do you cope with the conundrum posed by these two mutually opposed needs? One answer is to use mist propagation (see page 27) an ex-

Taking care of cuttings

Cuttings produce roots most rapidly under warm conditions, most successfully in bright light, most certainly when the atmosphere around the leaves is kept saturated with water

Cuttings enclosed inside a propoagator are easily killed by exposure to direct sunshine (*Right*) Out of the sun, inside a room, light is usually too dim for cuttings to do well

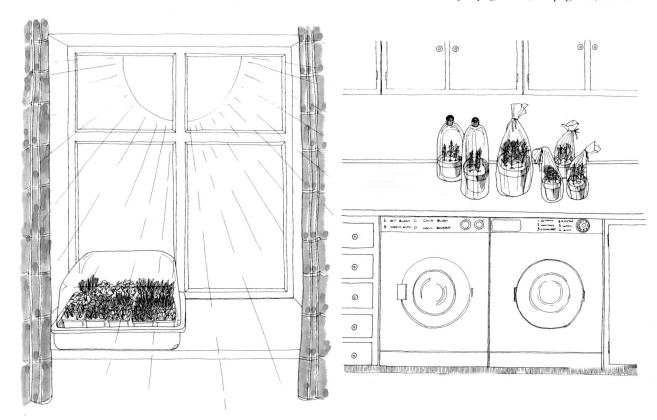

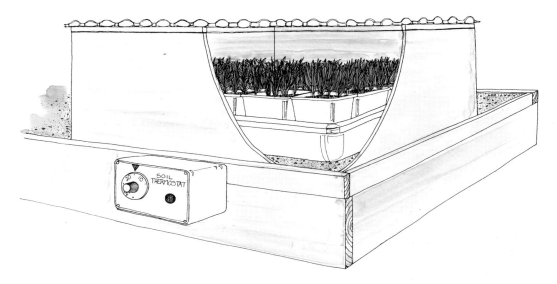

pensive solution that can bring almost as many problems as it solves. The alternatives depend on compromise:

1 Using just enough shading to prevent overheating, without reducing light levels too much;

2 Maintaining high humidity by sprays of water, bearing in mind that the time most of us have to do this is limited.

An ideal balance between light, warmth and humidity is provided inside a frame shaded with bubble polythene

Growth also depends on oxygen for cell division and the formation of tissues that make up new organs, roots among them. So cutting composts must hold air as well as water, and be sufficiently porous (open) for air to diffuse through them.

All this growth and development is affected by temperature. Things happen faster at higher temperatures and it is easy to assume that is a good thing. But the hardy plants dealt with in most of these recipes are naturally well adapted to grow in cool conditions. They may find a very warm atmosphere stressful, especially when, as cuttings, they are called upon to cope with new situations. Roots will develop more slowly at lower temperatures, but failures may be less likely, and a successful result is more important than saving time. High temperatures may hardly be avoided on hot days when the sun shines but these are naturally followed by lower night temperatures which bring relief. Artificial sources of heat that maintain constantly high temperatures are seldom beneficial when propagating plants from cuttings.

During the day, temperatures of 25°C (77°F) are more than enough, and extra shading or sprays of water should be used to prevent them rising higher. Most cuttings will respond well to working temperatures of around 15°C (59°F), and no harm will be done if up to half each day is spent at 10°C (50°F).

FIRST SHOOTS

Tip and Basal Cuttings in Spring

16

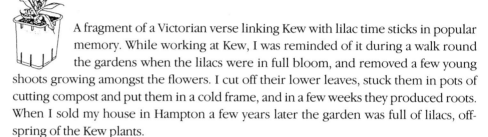

A fragment of a Victorian verse linking Kew with lilac time sticks in popular memory. While working at Kew, I was reminded of it during a walk round the gardens when the lilacs were in full bloom, and removed a few young shoots growing amongst the flowers. I cut off their lower leaves, stuck them in pots of cutting compost and put them in a cold frame, and in a few weeks they produced roots. When I sold my house in Hampton a few years later the garden was full of lilacs, off-spring of the Kew plants.

Lilac cuttings are likely to produce roots only at a particular stage of development, which happens to coincide with the week or so when the flowers are at their best. Unknowingly, I had taken those cuttings at exactly the right time. Only a few shrubs share lilac's preference for a restricted season. Many others that can also be grown from these early cuttings, but produce roots readily later in the summer too. Then it is better to wait if you can. Young shoots are fragile with little protection against adversity. They need more care than the tougher shoots of summer.

RECIPE 24	**Cuttings of 'Old English' Lavender**

Lavenders grow to precise schedules. By late spring every bush is covered with vigorous up-thrusting little shoots. A month or two later, depending on the variety, flower spikes appear at the tips of practically every one of these shoots. A visit to a lavender bush in search of cuttings in late spring/early summer produces masses of cuttings, and a few weeks later none at all – then a long wait till new shoots appear after the flowers are over and done with.

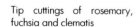

Tip cuttings of rosemary, fuchsia and clematis

Rosemary

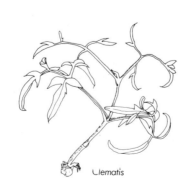

Clematis

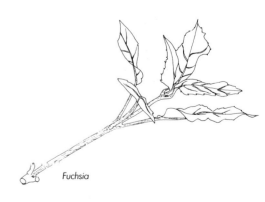

Fuchsia

An early start gives the young plants time to grow large by autumn. Cuttings can be taken from lavender bushes in the garden, but in chilly springs these will not be ready till late in the season: plants that have been forced a little in polythene tunnels are sold by most garden centres in the spring and are an excellent source of cuttings. One of these, sacrificed to participants on a Garden School propagation day was stripped of every shoot it possessed – seventy five in all. Every one a potential new plant!

Lavenders, and other plants with narrow silvery leaves, grow in hot, dry places. Even their immature shoots are better able than broad-leaved shrubs to resist drought – but they rot more easily if kept too wet. Cuttings of these plants should not be shut away too closely in shaded, water-saturated frames or propagators, but put where the warmth of the sun, and a little ventilation, preserve memories of their natural homes.

Tip cuttings are made from immature shoots soon after they begin to grow in spring. They are delicate and need to be looked after carefully. They can produce roots very rapidly under warm, light conditions – but are very vulnerable if they begin to dry out

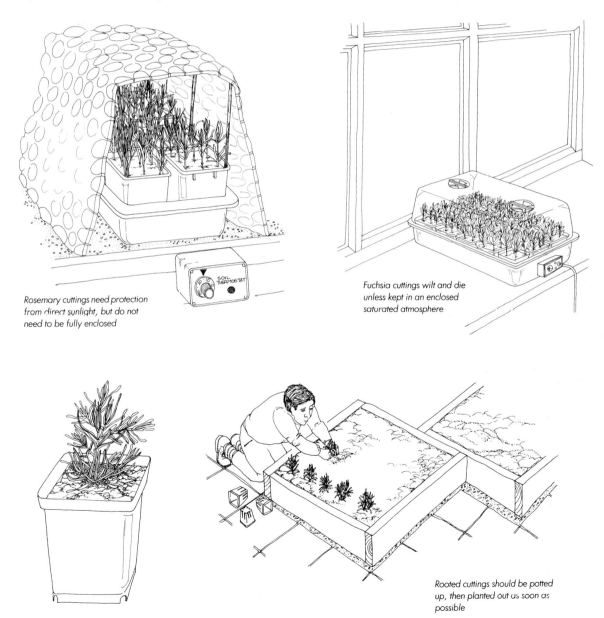

Rosemary cuttings need protection from direct sunlight, but do not need to be fully enclosed

Fuchsia cuttings wilt and die unless kept in an enclosed saturated atmosphere

Rooted cuttings should be potted up, then planted out as soon as possible

83

OTHER SHRUBS WHICH CAN BE GROWN FROM CUTTINGS IN A SIMILAR WAY

Artemisia *
Caryopteris
Ceratostigma
Cotton lavender
Helichrysum*
Osteospermum
Rock rose
Russian sage*
Thyme

*These are particularly vulnerable to overwatering and infection by water-borne fungi. Water thoroughly with tap water then allow the compost almost to become dry before watering again. Do not spray the foliage.

WHAT TO DO

1 Wait till the young shoots are 5 to 8cm (2–3in) long in late spring, just before flower spikes become visible.

2 Cut each shoot off at the point where it has started to grow. Find the vestiges of last year's leaves, the point at which they give way to the fresh growth of the current season is the place to make the cut.

3 Fill 7cm (2½in) square pots with a 50/50 mix of grit and perlite.

4 Remove the lower two or three pairs of leaves – nipping them off between finger and thumb – and dip the base of each into hormone rooting powder. Tap off excess powder.

5 Use a fine dibber to make holes in the compost and put nine cuttings into each pot.

6 Pack the pots into a seed tray under a bubble polythene canopy on a heated bench in a greenhouse. Water thoroughly. Use a thermometer to set the thermostat to maintain temperatures in the compost at around 15°C (59°F).

7 Check the cuttings every day. Water sparingly. In dull weather spray seldom and raise the sides of the canopy. In sunny weather spray twice a day. Close the sides of the canopy only if the cuttings start to wilt.

8 After three weeks drench the cuttings with a high-potash liquid feed (4×). Check for roots by pulling gently at one or two cuttings.

9 If roots are present, remove the pots of cuttings a week later and stand them on the greenhouse bench.

10 Two weeks later pot the rooted cuttings individually into a standard potting compost in 7cm (2½in) pots and put them back in the greenhouse.

11 After a month move them to a cold frame with the front permanently raised to provide ventilation.

12 Keep them in the frame, feeding with a high-potash liquid feed (1×) every fortnight. When the young plants are about 15cm (6in) high plant them in a nursery bed or out in the garden.

POSSIBLE PROBLEMS

SYMPTOMS	CAUSE	REMEDY
Lower parts of the cuttings turn black and die.	Infection by water-borne fungi. cf. damping off.	Water sparingly, and never let the compost become waterlogged. Use tap water until the cuttings have rooted.
Tips of the cuttings wilt, shoots shrivel.	Cutting compost has dried out, or atmosphere around cuttings has become too dry.	Always water before compost becomes very dry, and on hot, sunny days mist lightly.
Rooted cuttings fail to grow after potting up individually.	May be due to cold; more likely to starvation.	On cold nights and during cold spells keep framelights closed. Apply liquid feed every fortnight.

Clematis Cuttings in Early Spring

RECIPE 25

Almost the first promises of renewed growth every spring are the shoots that appear on clematis. And, if they are late-flowering varieties, every spring we prune them – cutting off and wasting those promising young growths. These early shoots are nothing like the long stemmed, woody ones that develop from them a few weeks later and make clematis such effective climbing plants. They are compact and leafy, looking more like something we would expect to find on a more regular shrub.

They need not be wasted. Used as cuttings, they are the easiest and quickest way to produce a new plant. Almost any climbing clematis will do, but one that I find particularly tempting to grow is the yellow *C. orientalis*, with thick petals like segments of kumquat peel. Cuttings taken in mid-spring produce roots so rapidly, and develop so fast that they can be nearly 2m (7ft) high and in flower by the autumn.

'Hagley Hybrid' is a pink clematis with an unusually long flowering season during the latter part of the summer. Pruning is uncomplicated: it can be cut back very hard each spring. Large numbers of promising young shoots will be discarded amongst the prunings, and some of these could be used to make cuttings. Clematis cuttings made from these early shoots (recipe 25) produce roots rapidly, and are a more reliable method of propagation than internodal cuttings made from the long stems later in the summer

———— WHAT TO DO ————

1 In early spring when the young shoots are 3 to 5cm (1½ to 2in) long – just before they start to elongate – cut off as many as you need at the point where they are attached to the main stem.

2 Remove the tiny leaves around the base, and all but one pair of fully developed leaves, nipping them off between your finger and thumb. Leave the bud and unopened leaves at the tip.

SHRUBS FROM EARLY SPRING CUTTINGS

Many shrubs can be propagated from immature shoots produced in the spring, but are easier to grow from cuttings taken a few weeks later when they are a little more mature. The following, however, are better grown from cuttings taken early in the season:

Abelia
Japanese maple
Lilac *
Smoke bush*

*These will not grow large enough during one season to be planted in permanent positions in the garden. Set them out 50cm (20in) apart in a nursery bed where they can grow on for another year.

3 Fill some 7cm (2½in) square pots with a 50/50 mixture of grit and perlite. Dip the base of each cutting into hormone rooting powder, tap off the excess and use a dibber to insert the cuttings about 1.5cm (½in) deep in the compost. (Nip out the terminal leaflets if they are in the way).

4 Put the cuttings in a frame on a greenhouse bench warmed with soil-heating cables.

5 Water the cuttings thoroughly and put on the framelight – making sure that it is a close fit. Adjust the thermostat to maintain a temperature of around 15°C (59°F) in the compost.

6 Keep the atmosphere around the cuttings permanently moist. In sunny weather spray once or twice a day with a cloud of fine water droplets – dampening the foliage but not adding much water to the compost in the pots.

7 When the cuttings start to produce roots, reduce the frequency of the sprays. One week later move them onto a nearby part of the bench, sheltered beneath a bubble polythene canopy.

8 Feed with a high-potash liquid feed (4×).

9 After three weeks pot the rooted cuttings individually into 7cm (2½in) square pots. Put them back on an open bench in the greenhouse.

10 Ten days later move the cuttings to a well-ventilated cold frame in the garden. Grow them on until they are 30cm (12in) high, tying each one to a short length of split bamboo cane. Then pot them into 12cm (4½in) pots.

11 Put them back in the frame and ten days later move them to a sheltered place outdoors. Push a 1m (3¼ft) bamboo cane into each pot and tie the plants to it as they develop.

12 When the plants reach the tops of the canes plant them in the garden. Do not line them out in a nursery bed.

Note: The large-flowered hybrids can be grown in the same way, but are not such fast movers. They should be kept in one litre pots until mid-autumn, or the following summer, before being planted in the garden. They will grow more rapidly if they are kept in the greenhouse and not moved into a cold frame or stood out of doors.

POSSIBLE	PROBLEMS	
SYMPTOMS	**CAUSE**	**REMEDY**
Cuttings wilt and fail to produce roots.	Dry atmosphere around the leaves of the cuttings.	High humidities are essential until roots develop on the cuttings. Spray with water frequently during the first ten days.
Plants fail to grow well after potting up.	Lack of nutrients leading to starvation.	Feed with a twice-standard concentration of high-potash liquid feed.

Border Phlox from Basal Shoots   *RECIPE 26*

I seldom need to resist an urge to make a herbaceous border; and if one starts to creep up on me I have found a perfect antidote. A moment spent pondering the demands of the tall, beautiful, fragrant border phloxes convinces me that I should continue to enjoy this particular piece of gardening over other people's fences – not within my own.

These gorgeous beauties thrive only on the best of treatment, apart from one or two resilient sorts which endure, with a suffering expression, neglect and poor conditions. They demand regular feeding, plenty of water, weed-free space, careful staking and renewal every few years, plus attention which I have no time to lavish, however rewarding these plants may be to those more prepared to make an effort.

Phloxes grow differently from other herbaceous plants. They form a matted foundation at and just below ground level from which the shoots appear every spring, and this cannot easily be broken up to make divisions. But the shoots which grow from it can be used as cuttings and produce roots with little difficulty. The basal shoots of many other herbaceous plants are very easy to strike (see the list on page 88), but most are even easier to grow from divisions, so cuttings are used only when there is some reason for doing so. One occasion which provides a reason is the purchase of a plant from a garden centre; a few cuttings taken when you get home can quadruple your stock at a stroke.

Cuttings of border phloxes made from the basal shoots.

Many herbaceous plants produce strong shoots from about ground level in the spring. These can be cut off just above their roots and make excellent, usually easily managed, cuttings. Like the border phloxes shown here, many do well in a cold frame.

———— WHAT TO DO ————

1 Take cuttings in mid-spring; but only from plants which grew strongly the previous season, with healthy leaves from top to bottom of their stems.

2 Cut the young shoots with a sharp knife below ground level where they are joined to the parent plant. Most will have a short, knobbly length of white stem at their base.

3 Fill 9cm (3½in) square pots with a standard potting compost.

4 Nip off the lower leaves of the cuttings between finger and thumb nail, leaving three or four of the upper leaves and the apical bud.

5 Dip the base of the cutting in hormone rooting powder, and use a pencil-sized dibber to insert each one 2 to 3cm (around 1in) into the compost. Each pot will hold nine cuttings.

6 Pack the pots into a seed tray, put it in a cold frame and water thoroughly.

7 Replace the framelight, making sure that it fits well. Lightly spray the cuttings once or more each day in sunny weather to maintain a humid atmosphere in the frame.

8 The cuttings will start to make fresh growth as soon as they have roots. Pull them very gently to check that they do have roots, and remove to an adjacent frame with the front of the light raised 10cm (4in). Close it only when cold winds are blowing or on chilly nights.

9 About two weeks later pot the cuttings individually into 7cm (2½in) pots using a general-purpose potting compost. Return them to the cold frame.

10 After about a month they should be ready to plant in the garden. Prepare planting spaces and set them out, marking and protecting each small plant with a short bamboo cane or sprig of brushwood.

Pot up the rooted cuttings individually as soon as the new roots are well developed, and move to a nursery bed a few weeks later

POSSIBLE PROBLEMS		
SYMPTOMS	**CAUSE**	**REMEDY**
Cuttings wilt and fail to form roots.	Atmosphere in frame too dry or too hot.	During hot sunny spells spray with water more frequently. Drape a sheet of milky polythene over the framelight as a diffuser.
Young plants fail to grow well. All their older leaves wither and die.	Stem eelworm* present in the shoots. This only affects parts of the plant above ground level.	Burn the cuttings. Cuttings made from the roots (see page 127) will be free from eelworms, but must be planted in clean ground.

*This is not likely to attack other plants in the list below.

OTHER PERENNIAL PLANTS (HERBACEOUS AND ALPINE) WHICH CAN BE PROPAGATED FROM BASAL CUTTINGS IN SPRING			
Achillea	Chrysanthemum	Geranium	Pink
Anthemis	Dahlia	Gypsophila	Shasta daisy
Aster	Dead nettle	Helenium	Thrift
Bergamot	Delphinium	Lupin	Valerian
Bugle	Diascia	Michaelmas daisy	Veronica
Campanula	Erigeron	Monkey flower	Viola

BONANZA TIME IN GARDENS FULL OF CUTTINGS

17

Falls following pride have brought me down to earth so often that even my stubborn nature has been forced to accept that the knack of learning how to make things easy is one of the great skills of gardening. Taking pride in overcoming problems, or triumphantly producing a batch of plants in spite of difficulties are fine. But there will be as many disasters as triumphs, and the vigilance and extra care needed make demands that any sensible person would avoid if possible.

Mid- to late summer is a time when everything combines to make the propagator's job easy. There are plenty of shoots to pick and choose from; they have become adjusted to the world around them and are no longer ultra-sensitive to heat, draughts or drought: they are growing actively and very ready to produce roots. Summer temperatures are high enough, night and day, to make artificial heat unnecessary, and the sun, though it may not appear as often as sunbathers would like, provides all the light that cuttings need to produce roots and grow well afterwards.

Making the Most of Plenty

Anyone who needs plants to fill a garden, and enjoys an easy life, should propagate while the going is good in summertime. This is such an outstanding season of opportunity that it is worth looking in a little more detail at the ways that well-chosen equipment and suitable methods can be used to ensure success.

It may not have escaped your notice that the time we are talking about is holiday time. Your family will not love you if you insist that the urge to propagate your alpines is more important than the annual breakaway. If you are going away – on holiday or to enjoy days out – then the alpines must be propagated in ways which make that possible. Nor should the neighbour, who comes in to open a can and put out a saucer of milk for the cat, be roped in for extra duties amongst the cuttings.

You can choose whether to go for methods which require daily attention for success, or to use others which are more or less self-propelled. The former can give quicker results and, if done well, may be more successful with a wider range of plants. But a popular way to cope with cuttings in days past was to stick them in a circle in a corner of the kitchen garden, cover them with a bell jar well daubed with whitewash, and leave them to it. Many of the more amenable shrubs and other plants, including alpines, produced roots within a month. Today those bell jars with the knobs on top are collector's pieces. Instead you can drape a short sheet of bubble polythene over two or three hoops in the ground to make a cloche, which will serve exactly the same purpose.

Successful gardening need not always depend on hard work and sweat.

Frames filled with rooted cuttings can be produced with constant care and attention, but equal success may result when simple precautions are followed, such as:
● covering frames with bubble polythene – not glass
● siting frames out of the full glare of the sun
● adding extra insulation – another sheet of bubble polythene or rush matting on sunny days
● covering the cuttings inside the frame with a thin sheet of polythene

The alternative to the bell jar for less work-shy gardeners used to be the sun frame. Usually made with brick sides and a glass-glazed framelight, it would face due south straight into the sun. The glass would be brushed with whitewash to temper the sun's strength, but success depended on keeping the atmosphere in the frame constantly saturated with fine sprays of water, applied – perhaps hourly on a hot day – with a garden syringe. Today a mist propagation unit (see page 27) can save us repeated forays with the syringe. But, when you depart on your holidays, leaving your cuttings dependant on mist propagation, you can be quite sure that it will go wrong within an hour of your departure and the cuttings will be dead long before you come home.

That risk can be avoided by sticking to the principle of the sun frame: siting and managing it according to the amount of time you can spare to look after it. The most effective ways to reduce its demands are as follows:

1 To use large cell bubble polythene sheets instead of glass. This diffuses sunlight, keeping the inside of the frame light but cool even on hot days.
2 To reduce exposure to the sun's heat either by siting the frame in a partially shaded postition, or by setting it up so that it faces north.
3 By adding extra shading (or insulation) during very hot weather. This can either be a second sheet of bubble polythene, or a sheet of milky white polythene, draped over the frame.
4 By reducing water loss by making a frame within the frame with a thin sheet of clear polythene on a wooden frame, just above the tops of the cuttings.

The more shading you use, and the cooler the interior of the frame, the longer it will take for the cuttings to form roots. In the end most of them will manage this as successfully as those that do it more quickly under warmer, more attention-needing conditions. A frame facing north, covered with a layer of bubble polythene and with an inner membrane of clear polythene, will look after itself for weeks at a time. A thorough drenching before departing on holiday will keep it going for three, four or even five weeks. In very hot sunny weather, drape another sheet of bubble polythene over the top to make sure!

How not to Kill Cuttings

People often tell me that their cuttings produce roots all right, but things begin to go wrong after that. It would not surprise me if a survey showed that more cuttings failed afterwards than die before they make a root at all. Aftercare is all important and some of the causes of failure are summarised below.

1 **Potting up cuttings too soon.**
   Make sure roots are well developed and that *all the cuttings that are likely to form roots have done so* before attempting to move them into pots.

2 **Potting rooted cuttings into unnecessarily large containers.**
   The smallest pot that will hold the roots without squashing them is the right one; move them into a larger one when they are ready.

3 **Potting up late in the season.**
   (This applies mainly to deciduous shrubs and plants, like heathers, with very fine roots). Rooted cuttings should always have time to grow and establish themselves before the onset of winter. If they are not ready to pot by the beginning of autumn, it is better to leave them in the cutting compost and feed them until the following spring, and then pot up.

4 **Exposing rooted cuttings and young plants to two stresses simultaneously.**
   Avoid potting up rooted cuttings, and moving them to different conditions at the same time. If they are to be moved, perhaps from a greenhouse to a cold frame, do this a week before, or a week after.

5 **Failing to saturate the compost with water after potting.**
   Peat composts, in particular, can look wet on the surface but remain dry beneath. Water thoroughly and check to make sure that the *compost around the roots* has been well wetted.

Cloche-Grown Mock Oranges — RECIPE 27

TAKE CUTTINGS: Mid- to late summer
INTO NURSERY: Late winter to mid-spring
PLANT: Mid- to late autumn

If you have a bell jar, this is the recipe when it can be used. If not, you can make an enclosed cloche by stretching bubble polythene over two or three steel hoops.

—— WHAT TO DO ——

1 Choose a sunny part of the garden – ideally a spot that is in partial shade around midday.

2 Mark out the position to be occupied by the cloche and rake it level. Top dress with 2cm (¾in) of a 50/50 mix of grit and perlite and fork it into the top 2 or 3cm (around 1in) of soil with a handfork.

3 Set up the hoops of the cloche and put the bubble polythene in place, folded back along one side.

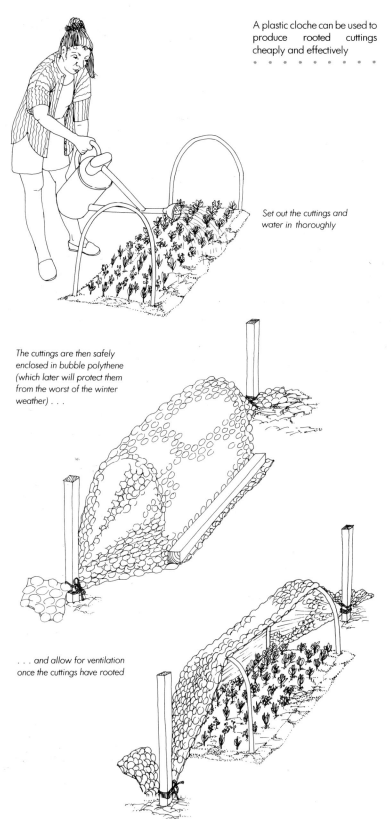

A plastic cloche can be used to produce rooted cuttings cheaply and effectively

Set out the cuttings and water in thoroughly

The cuttings are then safely enclosed in bubble polythene (which later will protect them from the worst of the winter weather) . . .

. . . and allow for ventilation once the cuttings have rooted

4 Take cuttings by pulling off young side shoots 7.5 to 10cm (3 to 4in) long. Do not use shoots with flowers. Each will have a little tag of tissue (a heel) torn from the main stem. Trim off the ragged part of the heel with a knife, and nip off all but the upper pair of leaves and the bud.

5 Dip the ends of the prepared cuttings into hormone rooting powder, tap off the excess, and set them out beneath the cloche in rows 5cm (2in) apart with about 5cm (2in) between each.

6 Water thoroughly and close the cloche, tying the polythene to a stake at both ends and holding the sides down with a little soil or a batten of wood. (Try to do enough cuttings of as many different plants as you want, to fill the cloche in one go.)

7 Check the cloche from time to time, if possible without opening it. It should not need watering but this must be done if the cuttings dry out and start to wilt. In hot, sunny weather drape a second sheet of bubble polythene over the top.

8 After four weeks open the cloche and pull gently at a few cuttings to test for roots. If present, raise one side of the polythene about 10cm (4in). Spray with water for a day or two in hot, dry weather, gradually raising the polythene along one side.

9 Leave the cuttings in place until spring; in very cold winter weather use the cloche to protect them from winds or frosts.

10 In early spring line out the rooted cuttings 30cm (12in) apart in a nursery bed; mulch well.

11 They can be planted in the garden during the following autumn.ˈ

POSSIBLE PROBLEMS

SYMPTOMS	CAUSE	REMEDY
Cuttings dry up and fail to form roots.	Interior of frame too hot and dry.	It may be difficult to keep the atmosphere in the frame humid during prolonged hot and sunny weather. Add extra polythene or splash with whitewash to reflect heat.
Rooted cuttings fail to establish in the nursery bed.	Lifted too soon; either before they are weaned, or their roots are sufficiently developed.	Be in no hurry to move the cuttings on. They will do better growing on under the cloche than planted out before they are ready.

Ground cover has become a familar phrase to gardeners. It conjures images of dwarf shrubs like heathers and spreading perennials — hostas perhaps? On a much smaller scale, many of the compact, ground-hugging plants we think of as alpines — like this alpine phlox — are useful weed excluders on the edges of paths, In small beds or along the tops of retaining walls. They can be propagated with ease during the summer (recipes 27, 28 or 29) from cuttings taken from newly grown shoots as soon after the flowers have faded as possible

Achillea	Dead nettle	Hypericum	Pink	Snow-in-summer
Alpine phlox	Diascia	Hyssop	Potentilla	Stonecrop
Box	Escallonia	Iberis	Privet	Tamarisk
Buddleia	Flowering currant	Kerria	Rock rose	Thrift
Bugle	Gypsophila	London pride	Senecio	Tree Mallow
Campanula	Hebe			

OTHER PLANTS THAT CAN BE GROWN FROM CUTTINGS UNDER A CLOCHE

| RECIPE 28 | **Hydrangea Cuttings in a Propagator Outdoors** |

TAKE CUTTINGS: Late summer
FEED: Mid-autumn and early spring
POT UP: Mid-spring
INTO NURSERY: Mid-summer
PLANT: Mid- to late autumn the following year

OTHER SHRUBS WHICH
CAN BE PROPAGATED
FROM CUTTINGS IN
A PROPAGATOR
OUTDOORS

Box
Buddleia
Deutzia
Escallonia
Euonymus
Exochorda
Firethorn
Flowering currant
Hebe
Hypericum
Iberis
Jasmine
Kerria
Mexican orange
   blossom
Mock orange
Periwinkle
Potentilla
Rock rose
Senecio
Shrubby honeysuckle
Spiraea
Tamarisk
Tree mallow
Weigela

Cuttings, you may have been told, are easy to root in pots covered with plastic bags on a window sill. But when I do it, I don't find it easy at all. The sun shines and leaves me with dead, boiled-in-the-bag cuttings. The plastic bag is never the right size – either too tight a fit or too big and floppy. It won't slip over the top of the pot or, during the manipulations, the pot tips over. When I take it off, some cuttings become enmeshed in a fold and are pulled out. A wire frame fashioned from a coat hanger is said to help a little by supporting the polythene bag.

Anyone who is clumsy like me, and finds these antics with plastic bags frustrating, can avoid them by buying a simple propagator: no need to burden yourself with an electrically heated one. Transparent plastic tops replace floppy bags, and fit neatly over trays which hold pots containing the cuttings.

This makes a cheap and simple scaled-down alternative to a cold frame. A standard seed tray holds fifteen small, 7cm (2½in) square pots; each pot holds from five to fifteen cuttings – depending on the plant and the size of its shoots. You can work out how many cuttings will fit into a tray, and would need to be very mean indeed to suggest that they wouldn't justify the small cost of the propagator.

———— WHAT TO DO ————

1 Collect cuttings in late summer. Use a sharp knife to cut off young, non-flowering shoots with three, possibly four, pairs of fully expanded leaves. Avoid spindly, or very fat shoots.

2 Fill 7cm (2½in) pots to within 0.5cm (¼in) of the rim, with a 50/50 mixture of grit and perlite.

3 Cut off the leaves, apart from the top pair and the bud above them, slicing through the stalk of the leaf where it joins the shoot.

4 Dip the base of each cutting into hormone rooting powder; tap off any surplus.

5 Use a pencil-sized stick as a dibber to set out five cuttings in each 7cm (2½in) square pot, sticking each one 3cm (1¼in) deep into the compost. Do enough cuttings at one time to fill the tray. They need not all be hydrangeas – any of the shrubs listed here could be used (and a good many others).

6 Pack the pots into the base, water thoroughly and fit the propagator lid over the top. Close the vents.

7 Put the propagator in a sheltered place, light but out of direct sunshine. A corner between a north and east or west wall might do, or under an evergreen shrub. If exposed to direct sunlight for an hour or two each day, a drape of bubble polythene makes an effective parasol.

8 Do not open the propagator; do not wipe off condensation; do not water except when really necessary: and, above all, *do not tweak out cuttings to see if they have any roots!*

94

9 Check the propagator every four or five days. If condensation thins out, or disappears from parts of its surface, open it and water throughly.

10 When the cuttings have produced roots they will begin to grow again, and new, fresh green shoots will appear. Check that roots have been formed by giving a few cuttings a gentle pull – if they don't come out, they have roots!

11 Transfer the pots to a propagator in a slightly sunnier place with the vents open all the time. Drench with a high-potash liquid feed (5×).

12 A fortnight later take pots of cuttings out of the propagator and stand them in a sheltered corner. Feed at monthly intervals with liquid feed (1×) till the leaves fall.

13 The following spring feed again (2×) when the first signs of growth are visible. A fortnight to a month later, pot them individually into 9cm (3½in) square plastic pots.

14 When their roots encircle the pots, plant out the young shrubs. Put them in a nursery bed, or, well-marked and well-protected, into the garden.

15 Plants in a nursery bed should stay there during the following winter and summer and be moved in late autumn, in the case of deciduous plants, after their leaves fall.

A propagator made from a seed tray with a transparent top is a convenient means of producing moderate numbers of cuttings. It must be carefully sited out of direct sunshine. For example it could be put:

● in the shadow of an ever-green shrub
● below a greenhouse bench
● at the foot of a north-facing wall

A standard seed tray holds fifteen small square 7cm (3in) pots. Each of these can be used to hold cuttings of a different plant, or one cutting after it has produced roots

Hydrangeas can be propagated during the winter from cuttings made from the leafless shoots (recipe 36). They are also amongst the many shrubs that can be grown from cuttings during the summer (recipe 28 or 29). 'Annabelle' is a comparatively recent introduction and intriguingly different from the more familiar lacecaps and mopheads. It is a form of *Hydrangea arborescens*. This never grows into a tree, as its name suggests, but remains a small, slender-stemmed shrub, which usually produces rather nondescript heads of flowers. 'Annabelle's' flowers look more like adornments that grace fair heads at Chelsea than exhibits on the stands

POSSIBLE PROBLEMS		
SYMPTOMS	CAUSE	REMEDY
Cuttings fail to form any roots.	Stressed by excessive heat, lack of water or waterlogging.	Make sure the propagator is never exposed to direct sunlight. Use presence/absence of condensation as a guide when watering. Avoid composts that do not drain freely.
Subterranean parts of cuttings turn black.	Blackleg – usually due to fungal infection such as damping off.	Use tap water for watering, rather than stored water in a tub or butt.
Rooted cuttings die during the winter.	Frost damage, accentuated when compost is very wet/waterlogged.	Protect cuttings during severe cold or wet spells. Make sure drainage holes in pots are not obstructed by leaves, algae etc.

Spiraea 'Gold Flame' Produced in a Cold Frame RECIPE 29

TAKE CUTTINGS: Mid-summer
POT UP: Late summer
FEED: Early spring to early summer
STAND OUTSIDE: Mid-spring
PLANT OUT: Early to mid-summer

Not many years ago *Spiraea* 'Gold Flame' was unknown; now we see its squat domes of bright apricot, tan and amber foliage everywhere. It is a reliable, easily grown plant, but young plants have the brightest foliage, and it is worth replacing older bushes from time to time. There is no need to buy new ones, as it is one of the easiest of all shrubs to grow from cuttings. 'Gold Flame' is an early starter, and its shoots are well developed by early summer, so cuttings taken then can be pushed on a bit, potted up before the winter and grown quickly the following summer to plant in the garden only a year after they were started.

I like to put them in a frame, although they would produce roots in a propagator as easily as the hydrangeas in the previous recipe. But the frame is warmer, more sheltered and speeds things up – in fact, unless the numbers of cuttings needed are very small indeed, I prefer it for almost all summer cuttings.

——— WHAT TO DO ———

1 Around mid-summer when the first flush of growth is over, young shoots begin to firm up. Hold them in your fingers, and bend them slightly. You will feel the resistance of the developing woody tissues in the shoots.

2 Pull young, non-flowering shoots off about 5cm (2in) long – neither very spindly nor gross – at the point where they join the main stem.

3 Fill 7cm (2½in) square pots with a 50/50 mixture of grit and perlite.

4 Nip off all but the top pair of fully developed leaves and the bud. Dip the base into hormone rooting powder and tap off any excess. Put nine cuttings in each pot, using a dibber to stick them 2cm (¾in) deep in the compost.

5 Pack the pots in seed trays in the frame. Water thoroughly, damping down parts of the frame not occupied by pots.

6 Replace the framelight. Once, or more times, each day depending on its situation and the weather, spray the inside of the frame with water. This makes maximum use of the warmth of the sun to speed up rooting. Less demanding ways to manage the frame are described earlier in this chapter (see page 90).

7 When the tips of the cuttings show signs of renewed growth, pull gently at one or two cuttings to check whether roots have been formed. If they have, move the cuttings to an adjacent frame with the front edge of the framelights raised 10cm (4in). Drench them with a high-potash liquid feed (5×).

8 Pot the rooted cuttings individually into 7cm (2½in) pots in late summer, and put them back in the ventilated cold frame.

9 Keep them in the frame through the winter, closing the ventilators only

on very cold nights, or during freezing winds.

10 Feed in early spring with high-potash liquid feed (2×), and about a month later feed again. Take the pots out of the frame and stand them outside in a sheltered place.

11 They will be large enough to plant in permanent positions in the garden in late summer.

POSSIBLE PROBLEMS

SYMPTOMS	CAUSE	REMEDY
Ends of the cuttings rot and shoots die.	Cutting compost kept too wet.	When spraying to maintain high humidities in the frame, it is important not to overwater the cutting compost in the pots. Use a very fine spray to dampen the leaves only.
Cuttings wilt and fail to grow roots.	Too dry and/or too hot inside the frame.	Make sure that the entire surface of the frame is damp after spraying and the leaves covered with droplets of water. Protection from over-heating is best provided by covering with another sheet of bubble polythene.
Cuttings die during the winter.	Killed by over-watering when the plants are not growing.	Water very sparingly from late autumn to late winter. During cold spells add extra insulation or covering to the frame.

OTHER SHRUBS THAT CAN BE PROPAGATED FROM CUTTINGS TAKEN IN MID- TO LATE SUMMER

Spiraea 'Gold Flame' is very hardy, and only exceptionally severe winters would kill it. Winter cold and damp would be a more serious danger to some of the other shrubs (marked with a*) in the list below.

Abelia*	Euonymus	Indigofera	Rosemary*
Artemisia*	Euphorbia	Jasmine	Salvia*
Aucuba	Exochorda	Jerusalem sage*	Senecio
Berberis	Firethorn	Kerria	Shrubby honeysuckle
Box	Flowering currant	Kolkwitzia	Skimmia
Broom	Forsythia	Lemon verbena*	Tamarisk
Buddleia	Fuchsia	Lithospermum	Tanacetum*
Ceanothus*	Genista	Mexican orange	Thyme
Cistus*	Hebe	blossom	Tree mallow
Cotoneaster	Helichrysum*	Mock orange	Vaccinium
Cotton lavender*	Honeysuckle	Myrtle*	Viburnum
Daisy bush*	Hydrangea	Osmanthus	Weigela
Deutzia	Hypericum	Periwinkle	Winter savory
Elaeagnus	Hyssop*	Potentilla	Zauschneria*
Escallonia	Iberis	Rock rose	

Geranium macrorrhizum: **Preparations for Ground Cover**   RECIPE 30

TAKE CUTTINGS: Mid- to late summer
FEED: Early autumn and early spring
POT UP: Mid-spring
STAND OUTSIDE: Late spring
PLANT: Early to mid-summer

Customers who test a nurseryman's patience include those who announce as they arrive that they've 'decided to give ground cover a try'. An hour later their dismay at the thought of the cost of five to ten plants to the square metre gets the better of them, and they depart, saying they are sure half a dozen will do for a bank 10m (33ft) by 2m (6½ft). You know they will be back one day just to let you know that 'ground cover doesn't work'!

Frustrating certainly, but they're people after my own heart. I could never bear to pay for the thousands of plants I need in my own garden. Yet finding ways to let plants form their own almost self-sufficient communities saves me hundreds of hours of uncongenial weeding and scraping every year. The answer, for those of us with other uses both for our time and our money, is to grow the ground covering plants we need for ourselves. There could be no better starting point than one of the forms of that superlative vegetable duvet, *Geranium macrorrhizum*.

———— WHAT TO DO ————

1 Wait until the flowers wither after mid summer.

2 Cut off shoots 5 to 10cm (2 to 4in) long – usually fairly close to the point where they started to grow earlier in the year.

3 Fill a seed tray with a 50/50 mixture of grit and perlite.

4 Nip off all lower leaves, leaving two or three upper ones and the bud.

5 Dip the bases of the cuttings in hormone rooting powder and, using a broad dibber, stick them in the compost so that they stand upright. A seed tray holds forty-five cuttings arranged in five rows of nine.

6 Stand the tray in a frame, water thoroughly and dampen the surfaces not occupied by plants. Replace the framelight.

7 Spray the cuttings and the surrounding frame every day – more than once on hot sunny days.

8 Watch out for renewed growth and pull gently at one or two cuttings to confirm they have roots. Then move them to an adjacent frame with the vents open 10cm (4in).

9 Drench with a high-potash liquid feed (5×). Leave the cuttings in their tray in the frame through the winter, watering just enough to stop them drying out. These and many of the plants listed with this recipe could be potted up during the autumn. But they survive better if left in cutting compost in trays, or pots, and take up less space at a time when protected accommodation is often in short supply.

10 In early spring feed again (2×) and three weeks later pot the cuttings individually into 7cm (2½in) square pots. Put them back in the frame.

11 Move them from the frame after a month. They will be large enough to plant out during the summer.

Propagating thymes and other mat-forming plants.

Many creeping plants form roots naturally if given a little encouragement. In early summer, work a mixture of 50/50 grit/peat (or substitute) well into the mat, leaving the tips of the shoots exposed. Five or six weeks later lift out ready-rooted individual shoots, or small shoot systems, cut them free and pot up individually

POSSIBLE PROBLEMS

SYMPTOMS	CAUSE	REMEDY
Bases of cuttings rot without forming roots.	Compost too wet.	Use tap water for the initial watering. Try to avoid adding much water to the compost when damping down. Repeat watering only as and when absolutely necessary.
Cuttings dry up and wilt in the frame.	Atmosphere too dry or compost dried out.	Always keep frames as full of cuttings as possible, and dampen down frequently with light sprays. Compost in the trays should not dry out before roots are formed if this is done properly.

OTHER PERENNIAL PLANTS THAT CAN BE PROPAGATED FROM CUTTINGS OF THEIR SHOOTS TAKEN DURING LATE SUMMER

Achillea	Catmint	London pride	Sempervivum
Alpine phlox	Cerastium	Penstemon	Stonecrop
Alyssum	Dead nettle	Perennial wallflower	Tanacetum
Anthemis	Diascia	Phlox	Thrift
Aubrieta	Euphorbia	Pink	Thyme
Bugle	Gypsophila	Saxifrage	Veronica
Campanula	Iberis		

CUTTINGS OF HEATHERS AND AZALEAS

18

Novice gardeners quickly learn to prick up their ears at mention of 'ericaceous plants' – one of gardening's awkwardly obscure phrases, but the passwords to a club. Heaths and heathers, rhododendrons, azaleas, pernettyas and the flamboyant pieris are cousins in the botanical family of the *Ericaceae*.

They and other shrubs, usually evergreen, often low-growing, frequently rounded, share a relationship – and the majority of them share a preference for acid soils – which makes them unsuitable for those whose gardens are founded on basic clays and limestones. This is a restriction which is disregarded at your peril – most probably by the failure of the plants – not only in the garden, but also when propagating them.

Notable exceptions are the winter-flowering heathers, especially the offspring of *Erica carnea*. The species grows naturally above limestone rocks on mountains bordering the Mediterranean Sea, and the numerous cultivars derived from it make themselves at home on similar soils elsewhere.

Erica 'Springwood White' from Cuttings — RECIPE 31

Speak of ground cover and well-drilled gardeners instantly think of heathers. In the right place, and used in the right way, heathers and many other ericaceous plants are invaluable allies in the battle against weeds. One of the most effective, and attractive, is a low, ground-carpeting form of *Erica carnea* called 'Springwood White'. For most of the year it forms a mat of green, akin to a lawn; in late winter and early spring its spikes of white flowers combine particularly well with the smaller spring bulbs.

The essence of success when planting cultivars of ling or heather as ground cover, is saturation planting. Sixteen plants to the square metre grow together and quickly form an impenetrable carpet – and, you might say, it would be cheaper to buy carpet! But these plants are easy to grow from cuttings – one reason why they are relatively cheap to buy. With plans laid well beforehand, a few well-grown plants recruited from a garden centre provide the starting point for an army of occupation a year later.

OTHER HEATHS THAT CAN BE PROPAGATED IN A SIMILAR WAY ARE LISTED BELOW

The time when the cuttings are ready varies depending on when the plants flower – the summer-flowering lings being later than winter-flowering heathers. None are lime-tolerant, apart from the tree heaths.

Bell heather
Cornish heath
Cross-leaved heath
Dorset heath
Ling
St Dabeoc's heath
Tree heath

———— WHAT TO DO ————

1 When planning to take cuttings from plants in your garden, cut the dead flowering heads off the parent plants in late spring with garden shears.

2 In late summer snip off newly grown shoots 3 to 5cm (1¼ to 2in) long with a pair of scissors or sharp knife.

3 Fill 7cm (2½in) square pots with a 50/50 mixture of horticultural vermiculite and grit. ['Springwood White' is a lime-tolerant winter-flowering heather. Cuttings of summer-flowering heathers are averse to lime, and it would be

Grow your own heather garden in a cold frame

Heathers should be clipped back after flowers have faded

Young shoots about 4cm (1½in) long make good cuttings. Sixteen fit into a 7cm (3in) pot

Pot up individually the following spring

Cuttings are protected by a cold frame in winter – close only during sub-zero temperatures

necessary to use a granite or millstone grit, which contains no free calcium, and to pot up the rooted cuttings in an ericaceous compost. In areas with calcium-loaded, hard tap water, the cuttings and young plants should be watered with clean rainwater.]

4 It is not necessary to remove any of the needle-like leaves. Simply dip the base of each cutting straight into hormone rooting powder and tap off the excess.

5 Use a fine dibber to set the cuttings out in the pots, putting twelve to sixteen cuttings in each.

6 Pack the pots into a seed tray and put them in a cold frame. Water thoroughly, dampening all exposed surfaces in the frame. Replace the framelight.

7 Open the frame as seldom as possible. In dull weather the cuttings should not need spraying. In hot sunny weather maintain a humid atmosphere by spraying once or twice a day. Water sparingly.

8 In early autumn drench the compost with a high potash liquid feed (3×), and raise the front of the frame about 10cm (4in) for ventilation.

9 Leave the cuttings in the frame through the winter, watering when essential to keep the compost moist. The front should be raised except when cold winds blow, and on frosty nights.

10 In early spring water again with a high-potash liquid feed (3×).

11 When fresh green growth appears on the cuttings, knock them out of their pots and pot them individually in 7cm (2½in) square pots. Return them to the cold frame, keeping them well ventilated and well watered.

12 During late summer or autumn, line the plants out 10cm (4in) apart in a

nursery bed, or plant them in the garden. Two weeks before doing so, fork lightly and rake in a dressing of a general fertiliser at 100g/sq m (3½oz/sq yd). Mulch the plants with a 2cm (¾in) deep layer of shredded bark.

13 Early the following spring, transplant the young heathers to their permanent positions in the garden.

POSSIBLE PROBLEMS

SYMPTOMS	CAUSE	REMEDY
Rooted cuttings go mouldy during the winter.	Attacked by grey mould.	Keep front of frames open to maintain good ventilation except in the coldest weather. Spray with captan or benylate.
Rooted cuttings cut off, or disappear during the winter.	Grazed by fieldmice.	Set traps, using natural baits like pumpkin seeds. Allow a cat access access around the frame, but use a net to exclude entry.

Japanese Azaleas from Cuttings   RECIPE 32

The blowsy-headed hybrid rhododendrons that overwhelm so many gardens on acid soils are neither quick nor easy to grow from cuttings. But the little evergreen Japanese azaleas that cover themselves with flowers during late spring are not difficult at all, and are very valuable small shrubs; most useful to anyone with a garden on the right kind of soil.

———— WHAT TO DO ————

1 In late summer hold the new shoots between finger and thumb and bend them from side to side. Take cuttings when you can feel the resistance of developing woody tissues.

2 Pull off shoots 3 to 5cm (1¼ to 2in) long at the point where they join the main stem. Each will come away with a sliver of wood from the main stem.

3 Fill 7cm (2½in) square pots with a 50/50 mixture of lime-free grit and perlite.

4 Nip off the lower two pairs of leaves, and any leaflets around the base. Trim off the sliver of wood and dip the ends of the cuttings into hormone rooting powder. Tap off the excess.

5 Use a dibber to insert the bases of the cuttings about 1cm (½in) deep into the compost. Each pot will hold six cuttings.

6 Pack the pots into a seed tray and stand the tray in a propagating frame on a greenhouse bench above soil-heating cables. Water thoroughly and replace the framelight, making sure that it forms a close seal with the frame.

7 Use a thermometer to set the thermostat controlling the cables so as to maintain around 15°C (59°F) in the compost around the cuttings.

8 During hot, sunny spells drape an extra sheet of bubble polythene over the frame to prevent temperatures going above 25°C (77°F).

OTHER SHRUBS THAT CAN BE PROPAGATED IN A SIMILAR WAY

Cassiope
Enkianthus
Gaultheria
Kalmia
Ledum
Leucothoe
Phyllodoce
Pieris
Rhododendron
Vaccinium

When no greenhouse is available, cuttings of all these could be rooted in a cold frame in a similar way to the heaths described in the previous recipe.

Unlike many rhododendrons, the evergreen japanese azaleas are not difficult to grow from cuttings (recipe 32). They grow into low, often rounded shrubs covered with myriads of flowers. Many of the cultivars clothe themselves in brilliant shades of pink, magenta, cerise and crimson, guaranteed to liven up woodland settings of birches and bluebells. The white 'Palestrina' introduces a cooler, more harmonious note, and is a plant which can be used over and over again — making a virtue of repetition

unheated part of the bench in the greenhouse.

12 In early spring repeat the high-potash liquid feed (3×) and move the pots to a well-ventilated cold frame.

13 When young shoots start to develop, pot the rooted cuttings individually into ericaceous compost in 7cm (2½in) pots. Put them back in the ventilated cold frame.

14 During mid-summer prepare a nursery bed in a sheltered part of the garden. Add a 5cm (2in) top-dressing of 50/50 lime-free grit and composted bark to heavy soils and fork it into the surface layers. On light soils use composted bark only. Rake in a 100g/sq m (3½oz/sq yd) dressing of a general fertiliser sprinkled over the surface.

9 Spray the leaves lightly to maintain humidity, but disturb the atmosphere in the frame as little as possible. Do not overwater the compost.

10 In mid-autumn raise the front of the frame 10cm (4in), remove supplementary shading, and drench with a high-potash liquid feed (3×).

11 Remove the pots from the frame in late autumn and stand them on an

15 Two weeks later line out the young plants 15cm (6in) apart in the rows, water well if dry, and mulch with a 2cm (¾in) layer of composted bark.

16 Plant the young shrubs in the garden when they are big enough. This will probably not be until at least fifteen months after moving them to the nursery bed. Keep them scrupulously weed-free meantime.

POSSIBLE PROBLEMS

SYMPTOMS	CAUSE	REMEDY
Young leaves pale green/yellow. May shrivel at edges.	Chlorosis due to the effects of lime, either in the compost or in the water supply.	Check that all grits used are lime-free. Pot up in ericaceous potting compost. In hard water areas use rainwater not tap water.
Ends of cuttings go black. Fail to form roots.	Infection with water-borne fungi, such as damping off. Probably due to contaminated water.	Thoroughly clean and disinfect tanks used to store water at least once a year.

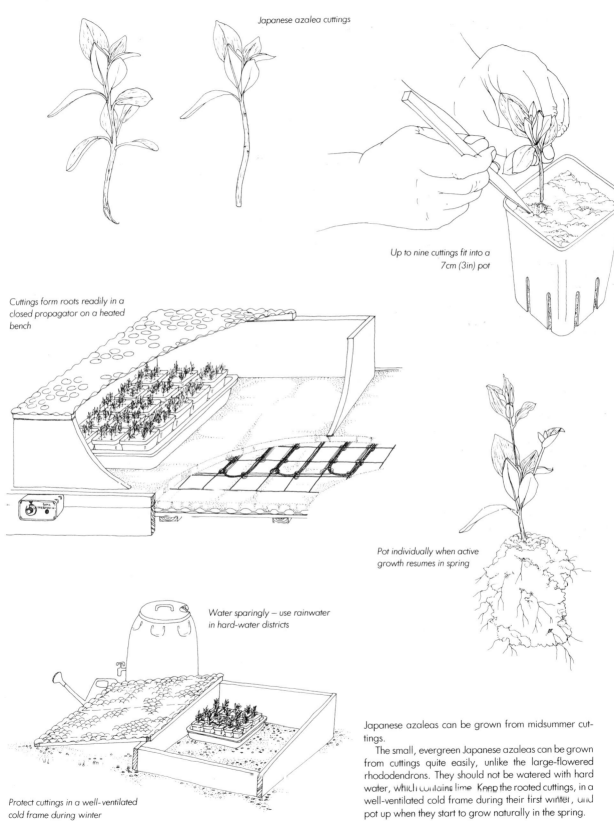

Japanese azalea cuttings

Up to nine cuttings fit into a
7cm (3in) pot

Cuttings form roots readily in a
closed propagator on a heated
bench

Pot individually when active
growth resumes in spring

Water sparingly – use rainwater
in hard-water districts

Protect cuttings in a well-ventilated
cold frame during winter

Japanese azaleas can be grown from midsummer cuttings.

The small, evergreen Japanese azaleas can be grown
from cuttings quite easily, unlike the large-flowered
rhododendrons. They should not be watered with hard
water, which contains lime. Keep the rooted cuttings, in a
well-ventilated cold frame during their first winter, and
pot up when they start to grow naturally in the spring.

105

19 INSURING TENDER THINGS

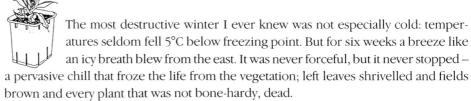

The most destructive winter I ever knew was not especially cold: temperatures seldom fell 5°C below freezing point. But for six weeks a breeze like an icy breath blew from the east. It was never forceful, but it never stopped – a pervasive chill that froze the life from the vegetation; left leaves shrivelled and fields brown and every plant that was not bone-hardy, dead.

Nonetheless, the desolation had a bright side. For once, I had gone round the garden during the autumn and taken cuttings of things I feared I might lose in a cold winter. When spring came, my greenhouses and frames were full of young hebes, penstemons, violas, fuchsias, sun roses and rosemarys, ready to replace those that were lost.

RECIPE 33 — Ceanothus Protected in a Greenhouse

Cold winters remind us that many attractive, evergreen shrubs are naturally attuned to warmer climates than ours. Gardeners learn from their losses that these are not long-lived, reliably hardy shrubs, and ceanothus sales drop with the temperature. But consumer resistance seldom lasts long. Ceanothus are such excellent, beautiful wall shrubs, and grow so quickly from replacements that most people are glad to give them a second, even a third, chance.

Severe winters will kill or damage many evergreen shrubs in the garden. The season when cuttings from them form roots most easily is in early spring, just before growth starts again. But after a cold winter the plants may be dead by then. Cuttings can also be taken in early winter; more will fail and they will take longer to form roots, but anything is better than nothing and this can be a useful insurance policy.

OTHER EVERGREEN SHRUBS WHICH CAN BE PROPAGATED IN A SIMILAR WAY.

Most of the shrubs listed below could be planted into a nursery bed and grown on for a further year before planting out.

Azara
Bay
Box
Cistus
Escallonia
Griselinia
Hebe
Holly
Japanese laurel
Mexican orange
    blossom
Myrtle
Pittosporum
Senecio
Skimmia

———— WHAT TO DO ————

1 In early winter cut side shoots off at the places where they join the main stem. Each should be 10 to 15cm (4 to 6in) long.

2 Fill 7cm (2½in) square pots with a 50/50 mixture of grit and perlite.

3 Nip the leaves off the lower half of each cutting; dip its base into hormone rooting powder and use a fine dibber to stick it 3cm (1¼in) into the compost. Each pot will hold nine cuttings.

4 Pack the pots rim to rim on a heated bench in a greenhouse. Shelter them beneath a canopy of bubble polythene on a light wooden frame.

5 Water thoroughly and damp surrounding surfaces. Set the thermostat for a bench temperature of around 15°C (59°F).

6 Check daily. Spray lightly to dampen the foliage on bright sunny days; water sparingly – enough to prevent the compost drying out.

7 After eight weeks drench with a high-potash liquid feed (5×). Continue to feed (1×) at monthly intervals.

8 In mid-spring pot rooted cuttings individually into 7cm (2½in) square pots. Place on an unheated part of the greenhouse bench.

9 Six weeks later pot the plants into one litre pots and stand them out in a cold frame. Ceanothus plants are put into one litre pots because they do not transplant well. Other evergreen shrubs that could be propagated similarly are listed below: most of these could be planted into a nursery bed and grown on for a further year before planting out. Keep the framelight closed at night and on all but bright, sunny days for one week; then raise the front to provide continuous ventilation.

10 Move the plants into a sheltered place out of doors for the summer. Plant against a wall when needed.

Economical ways to protect semi-tender plants in cold winters.

Heating a greenhouse to prevent cuttings and tender plants dying in the winter can be very expensive. Effective but cheaper ways are available.

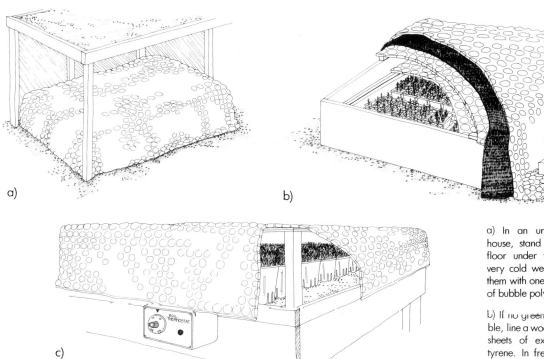

a)

b)

c)

a) In an unheated greenhouse, stand cuttings on the floor under the benches in very cold weather, covering them with one or more layers of bubble polythene

b) If no greenhouse is available, line a wooden frame with sheets of expanded polystyrene. In freezing weather, cover the cuttings in the frame with a sheet of bubble polythene. Add extra layers of bubble polythene over the top and sandwich an old carpet between them. Use these additional coverings only during very cold weather

c) Stand the cuttings on a heated bench and set the thermostat at 5°C. In freezing conditions, drape bubble polythene over the plants

POSSIBLE PROBLEMS		
SYMPTOMS	**CAUSE**	**REMEDY**
Buried stems of cuttings go black and fail to form roots.	Stem rot fungi, cf damping off.	Avoid overwatering the compost. Water with tap water – not water stored in a tank.
Cuttings stay healthy but produce no roots.	Ineffective rooting hormone.	Replace stocks of rooting hormones annually. Store in a dark, cool place, preferably in a refrigerator.

RECIPE 34 — Autumn Cuttings of *Penstemon* 'Firebird'

Penstemon popularity rises and falls like a yoyo. Not long ago they were underrated because everybody agreed they were tender. Now the fashion is to emphasise their hardiness – even when selling varieties which will be dead by Christmas in an average winter. *Caveat emptor*!

A few, amongst them 'Garnet' and 'Firebird'*, have persuaded me that they can stay alive through almost any winter – others range from surviving some to surviving none. Whatever their rating in the hardiness stakes, I like to take a few cuttings each autumn and overwinter them in a greenhouse in case disaster strikes. Even when it does not, the young plants are very welcome and homes are easily found for them, either in my own garden or in someone else's.

I always use a greenhouse with a heated bench for these and similar cuttings. A cold frame in a sheltered, sunny position would do, but cold winters would cause losses and the plants would not grow so well.

*Note that The Hardy Plant Society's annual compilation *The Plant Finder* gives the correct names for *Penstemon* 'Garnet' and 'Firebird' as 'Andenken an Friedrich Hahn' and 'Schoenholzeri', respectively. It is not only Latin names that are sent to try gardeners!

Plants that are not quite hardy enough to survive very cold winters should be propagated in the autumn as an insurance.

Cuttings on a heated bench in a greenhouse produce roots most rapidly, but seldom need to be fully enclosed. A canopy of bubble polythene, open at either end, provides all the protection they need

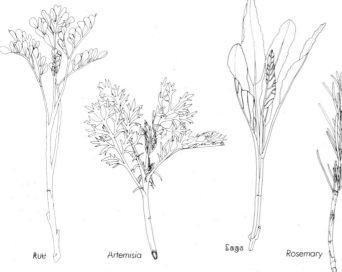

Rue   Artemisia   Sage   Rosemary

WHAT TO DO

1 Construct a canopy of bubble polythene on a wooden framework above a greenhouse bench warmed by soil-heating cables.

2 Take cuttings in early autumn by pulling off short – 7.5 to 10cm (3 to 4in) – non-flowering side shoots from the stems below the flowers.

3 Trim slivers of stem attached to the base of the cuttings with a sharp knife. Nip off two or three pairs of the older leaves.

4 Fill 7cm (2½in) square pots with a 50/50 mix of grit and perlite.

5 Dip the ends of the cuttings in hormone rooting powder, tap off the excess, and use a dibber to insert the cuttings about 2cm (¾in) deep in the pots.

6 Pack the pots on the bench beneath the canopy, and adjust the thermostat to maintain a bench temperature of 15°C (59°F).

7 Water the cuttings thoroughly. For

the first three or four days spray them lightly to dampen their foliage. After that they should not need to be sprayed to stop them wilting unless the weather is very hot and sunny. Water when necessary to keep the compost moist.

8 After four weeks drench the pots containing the cuttings with a high-potash liquid feed (3×). Remove the bubble polythene canopy.

9 Lower the thermostat to maintain bench temperatures of 5 or 6°C (41 or 43°F) – just enough to keep the cuttings frost free. On cold nights lay a sheet of bubble polythene over the rooted cuttings, removing it each morning unless the weather is very cold. Give monthly high-potash liquid feeds (1×) throughout the winter.

10 Pot the rooted cuttings individually into 7cm (2½in) pots in early spring. Return them to an unheated bench in the greenhouse.

11 A month later move the plants to a cold frame. For the first fortnight keep them closed except on warm,

sunny days. Then ventilate by raising the front 20cm (8in), except during periods of cold winds or on frosty nights.

12 Grow the young plants on in the frame, giving progressively more ventilation. Do not plant them in the garden till early summer.

The white form of *Fuchsia magellanica* scarcely qualifies as a tender plant. But sooner or later a dreadful winter will deal it a lethal blow, and it is one among many more or less tender plants that can be grown from autumn cuttings (recipe 34) to make sure it is not irretrievably lost when that dreaded day comes. Faintly flushed with pink, not white, its graceful flowers and attractively fresh green foliage really come into their own in the autumn – thriving during the dampest, mistiest and gloomiest days

Leave *Osteospermum* 'Butter-milk' out to take its chance one winter and you will be left in no doubt that this not a hardy plant. More often than not it will be dead by Christmas. It grows readily from cuttings and these can either be taken in the autumn (recipe 34) as an insurance against loss, or a stock plant or two can be bought in the spring and used (recipe 46) to raise a batch of plants that will be in flower during summer

POSSIBLE PROBLEMS

SYMPTOMS	CAUSE	REMEDY
Cuttings wilt and fail to form roots.	Atmosphere or cutting compost too dry.	During the first few days the cuttings need to be sprayed to prevent wilting. Later they should stand up without help.
Rooted cuttings die during the winter.	Over-watering accentuated by low temperatures.	Throughout the winter water sparingly. Even if the plants wilt briefly they will survive better kept dry than wet.

OTHER PLANTS WHICH CAN BE PROPAGATED FROM CUTTINGS TAKEN IN EARLY AUTUMN

Artemisia*	Fuchsia	Helichrysum*	Perennial wallflower*	Rue*
Diascia	Gazania*	Marguerite	Pittosporum	Sun rose
Felicia	Hebe	Osteospermum*	Rosemary*	Tanacetum*

*These are naturally adapted to withstand the effects of drought – but correspondingly are sensitive to damp. They do not need an overhead canopy, and should be sprayed sparingly, even on hot, sunny days, and preferably not at all. Take extra care not to overwater during the winter.

RECIPE 35   Late Cuttings of Violas

It is not easy to endow a garden with a sense of humour, but violas can! Pansies' 'faces' often scowl; violas show a sunnier disposition, and a mixed group can make me smile even on a poor day. They have other charms. Many are easy to grow. They flower all summer if encouraged by occasional feeds. Between them they cover a remarkable range of colours and combinations, from the subtlest pastel shades to the richest velvety depths.

Many of them, left to their own devices, make excuses and depart after a year or two, and time is well spent raising a few each year to keep the supply going. Late-struck cuttings form roots readily, and can be relied on to survive the winter in a cold frame. Penstemons might; violas will!

LATE CUTTINGS OF SHRUBS

The shrubs listed in the previous recipe can be brought through the winter in a sunny cold frame in most years. They are more vulnerable to frost than violas, and need extra protection in cold weather. Those that were marked with an asterisk are more liable to damp off. The following can usually be raised with no more difficulty than violas:

Diascia
Grevillea
Hardy fuchsia
Hebe

———— WHAT TO DO ————

1 Prepare the plants in late summer by cutting the flowering shoots off just above ground level.

2 In mid-autumn take cuttings from the newly developed shoots. These will be 2 to 4cm (¾ to 1½in) long, and should be cut off where they emerge from the centre of the plant.

3 Fill 7cm (2½in) pots with a 50/50 mixture of grit and perlite.

4 Nip the older leaves off the cuttings to form a short length of clear stem. Dip the base in hormone rooting powder, and use a dibber to insert the cuttings 1cm (½in) deep in the compost, six cuttings to a pot.

5 Pack the pots into seed trays and stand them in a cold frame. Water thoroughly, dampen down all surfaces within the frame. Replace the framelight, making sure it makes a snug fit.

110

6 Check the cuttings two or three times a week. In sunny weather provide extra protection for newly taken cuttings with a spare sheet of bubble polythene. Spray lightly from time to time.

7 After a month drench with a high-potash liquid feed (3×). Open the framelight about 10cm (4in).

8 Ventilate continuously throughout the winter, except during sharp frosts. Water sparingly – possibly hardly at all – depending on the siting of the frame. In very severe weather, add extra insulation to the cold frame.

9 In late winter feed (1×) again and about two weeks later pot the cuttings up individually into 7cm (2½in) pots. Return them to the frame.

10 Grow the plants on in the cold frame, with plenty of ventilation, till they are large enough to plant in the garden – any time from late spring onwards.

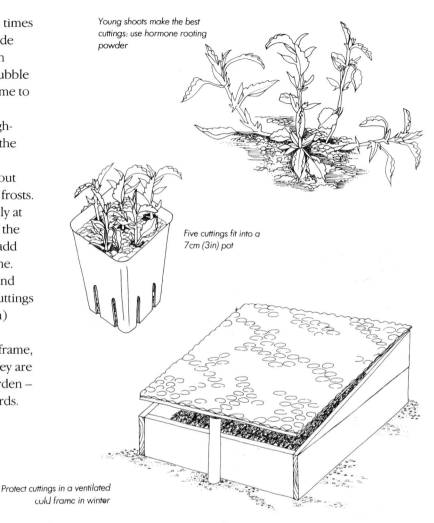

Young shoots make the best cuttings: use hormone rooting powder

Five cuttings fit into a 7cm (3in) pot

Protect cuttings in a ventilated cold frame in winter

P O S S I B L E P R O B L E M S		
SYMPTOMS	CAUSE	REMEDY
Plants die during the winter.	Damping off due to infection by fungi.	Site the frame where it gets as much winter sunshine as possible. Water Water with tap water and water sparingly during winter, not at all on dull days. Maintain free ventilation.
Plants disappear during the winter.	Grazed by mice.	Set traps baited with natural bait, eg sunflower seeds. Allow cats access around, but not in the frames.

Keeping up a supply of violas. Some of the most attractive violas do not set seed and are short lived. A few cuttings taken in early winter guarantees a good display the following summer. Cut back the old flowering shoots in late summer to encourage the plants to produce the young shoots needed for good cuttings.

WINTER CUTTINGS OF SHRUBS AND TREES

20

No plants are easier from cuttings than willows. Almost any twig or branch can be cut off during the winter (recipe 36) and stuck into the ground. Clusters of cells – preformed to develop into roots – guarantee roots by springtime, and the plants will take hold and grow through the summer as though they had always been a fixture. Willows have been branded as obstructors of drains and underminers of foundations, but there are dozens to choose from, many too small to produce roots that would block anything more than a metre or two away

I was sowing lettuces the first time I propagated blackcurrants. Looking around for something to mark the ends of the rows, I found some blackcurrant prunings, cut them into short lengths and stuck them in. By the time the lettuces were ready to eat, the prunings had produced roots and shoots and were growing into promising little blackcurrant bushes.

Those prunings were hardwood cuttings. Possibly the easiest, certainly the least demanding method of propagation anyone could hope to find. Some trees and shrubs, like willows and buddleias, produce roots so easily that every one is a winner. Others, and these include the red-stemmed dogwoods, are less obliging – perhaps a third succeed in a good year – but the effort is still well worth it.

RECIPE 36   **Roses from Hardwood Cuttings**

> TAKE CUTTINGS: Late autumn
> FEED: Early spring and late summer
> PLANT: Early winter
> PRUNE HARD: Early spring (year two)
> FIRST FLOWERS: Mid- to late summer

The roses you buy are mostly composite plants – put together from two different roses. The shoots with their flowers and leaves will be a cultivar with a name, and the label will tell you when it flowers, what colour it is, and how sensually it is perfumed. The roots will be seedlings, grown from a selected form of a wild briar rose, and the label won't mention them at all!

These roses are constructed by slipping a single bud of the cultivar into a slit made in the bark of the briar seedling just above the roots. The bud grows and, cuckoo-like, re-

112

places the briar's shoots. Unfortunately, the roots of some of the wild roses used produce suckers, and their offspring are unwelcome intrusions in anyone's rose beds.

Problems with suckers can be avoided by growing roses on their own roots, from cuttings. Any that appear are then part of the rose itself and can be left where they are. Nurseries seldom grow roses this way – for various reasons – but it is a practical way to produce a home-made rose.

Roses from cuttings taken as winter starts.

The best time to take these cuttings is as the leaves change colour, immediately before they fall, but cuttings made from leafless shoots also do well

———— WHAT TO DO ————

1 Choose a sheltered corner with well-drained soil and as much sun in winter as possible.

2 Dig out V-shaped trenches about 30cm (12in) apart and 10cm (4in) deep by 10cm (4in) wide at the top.

3 In late autumn cut off long straight shoots produced by the roses during the summer.

4 Cut each into sections to make several cuttings, with four buds on each. Dip the base of the cuttings into hormone rooting liquid.

5 Line the cuttings out along the trenches. Space them 15cm (6in) apart, with their uppermost buds about 2cm (¾in) above the surface of the ground. A square metre of the bed will hold about thirty cuttings.

6 Refill the trenches with a mixture of soil/grit/perlite at a ratio of 2/1/1.

7 Early the following spring, remove weeds; top dress the bed between the rows with 100g/sq m (3½oz/ sq yd) of general fertiliser; add a 3cm (1¼in) mulch of bark chips.

8 If necessary water during the summer, and from late summer onwards foliar feed every two weeks with a high-potash liquid feed (1×).

9 When the leaves fall, dig up the rose bushes and plant them in the garden.

10 The following spring, prune the shoots hard: cutting them off just above a bud 10cm (4in) above ground level.

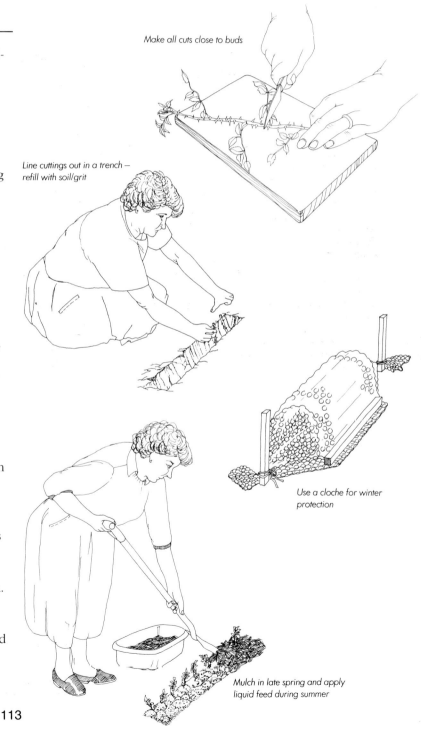

Make all cuts close to buds

Line cuttings out in a trench – refill with soil/grit

Use a cloche for winter protection

Mulch in late spring and apply liquid feed during summer

POSSIBLE PROBLEMS

SYMPTOMS	CAUSE	REMEDY
Cuttings fail to produce roots.	Situation too wet or winter too severe.	Avoid any situation which is waterlogged, even for short periods. During very severe weather cover rows with plastic cloches.
Leaves of young plants develop dark spots and fall prematurely.	Infection with black spot, a fungal disease that attacks some cultivars more severely than others (only affects roses).	Dig up and burn affected plants. Use other, more resistant cultivars next time.

OTHER DECIDUOUS SHRUBS THAT CAN BE PROPAGATED FROM HARDWOOD CUTTINGS

Plants you buy of these, unlike roses, will almost always be grown on their own roots:

Buddleia	Elder	Hypericum	Poplar	Tree mallow
Ceanothus	Flowering currant	Jasmine	Privet	Weigela
(deciduous)	Forsythia	Mock orange	Spiraea	Willow
Deutzia	Hydrangea	Mulberry	Tamarisk	Wisteria
Dogwood				

RECIPE 37 — Rapid Results with Ivies

TAKE CUTTINGS: Mid-autumn

FEED: Early winter

POT UP: Late winter

TRANSFER TO FRAME: Mid-spring

PLANT: Mid- to late summer

Some people talk as though they would be happier to share their homes with the Boston strangler than plant ivy in their gardens. I don't understand the prejudice. No other plants provide better, brighter and more varied winter evergreens, or are so tolerant of being overshadowed and forgotten during the summer.

Although rampant by reputation, *large* ivies, newly planted, can be slow starters. However, they are extremely easy to propagate, and have the engaging habit of growing through the winter: an early batch of cuttings will grow into small plants by late spring, ready to make themselves at home in the garden and fulfil their rampant destiny more quickly than would larger plants.

I have never come across any problems propagating ivies, but can anything in gardening be that foolproof? Let me know when you get tripped up!

Periwinkles respond equally amenably to similar treatment.

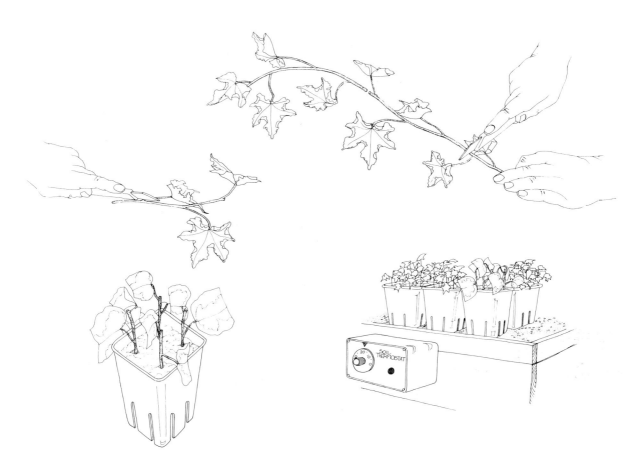

WHAT TO DO

1 In mid-autumn cut off long, trailing shoots that have grown during the summer, and slice them into sections with three or four leaves on each.

2 Fill 7cm (2½in) square pots with a 50/50 mixture of grit and perlite.

3 Remove the bottom leaf from each cutting. Dip their bases in hormone rooting powder and stick them 2cm (¾in) deep into the compost. Four to nine will fit into each pot, depending on how large their leaves are. Cut large leaves in half to reduce congestion.

4 Pack the pots into seed trays and put them in a cold frame. Water thoroughly, damp down the inside of the frame and replace the framelight.

5 Check the pots twice a week; water when necessary to prevent the compost

becoming dry. After six weeks feed with a thorough drench of high-potash liquid feed (5×).

6 In mid-winter move the cuttings into a greenhouse. Stand them on a heated bench if there is space, or put them on the floor beneath the benches. (Note that this is worth doing only if some of the benches in the greenhouse are fitted with functioning soil-heating cables. The cuttings can be left to grow on in the cold frame but will not develop so rapidly.)

7 In late winter pot the rooted cuttings individually into 7cm (2½in) pots and put them back into the greenhouse.

8 Plant them in the garden during the summer – marking their positions and protecting them, if necessary, with twigs of brushwood.

Ivies from cuttings: a useful source of winter greenery.

Trailing ivy stems are cut into short lengths, each with three or four leaves. Rooted cuttings grow well during the winter if stood on a heated bench. Large leaves can be cut in half to make them easier to fit into small pots

21 GROW-YOUR-OWN CONIFERS

Junipers were too flat, but 'Skyrocket' was too upright; one kind of Lawson's cypress was too green, another too dull; yet another more 'brassy' than gold; pines were 'ugly', and spruces reminded her of Father Christmas, but at last my customer decided that a neat little plant of *Thuja* 'Rheingold' was what she wanted. She asked, 'How do you propagate it?' 'From cuttings', I replied. 'Oh well then, I'll get a cutting from my sister's plant and save the money'. End of sale!

RECIPE 38 — **Cuttings of *Thuja* 'Rheingold' in a Cold Frame**

The little tree that the customer finally fell for might have been small, but it was over four years old, and it was the wrong time of year to take a cutting. Many conifers are not hard to propagate, but, in their early years, rooted cuttings develop steadily rather than rapidly. They need to be looked after carefully to grow into attractively shaped plants. They'll try your patience and test your skill: a few days inattention during hot weather, or failure to pot them up when they are ready, results in spavined, misshapen plants and, unlike a flowering shrub, a little judicious pruning will not restore shape and character. I have nearly always preferred to buy conifers when I needed them, ready made – and save my attempts at propagation for other things.

——— WHAT TO DO ———

1 Take cuttings during late autumn, removing shoots 5cm (2in) long which have grown during the summer. (Shoots towards the base of this cultivar often have more finely divided foliage than those produced higher up. Cuttings from the former will produce slower growing, more compact, rounded bushes.)

Conifers are not difficult to grow from cuttings, but patience and care are needed for good results·

The side shoots of conifers, used as cuttings, vary greatly in size. There is no reason to think that small cuttings form roots more readily than large ones or vice versa. Large cuttings grow more quickly into good sized plants, and should be used when a few plants are needed as quickly as possible

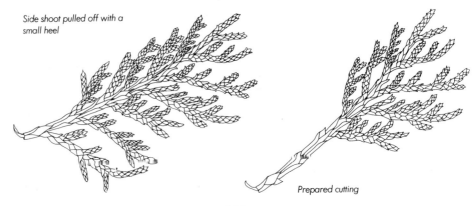

Side shoot pulled off with a small heel

Prepared cutting

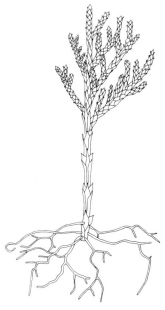

Rooted cutting ready to be potted

2 Fill 7cm (2½in) pots with a 50/50 mixture of grit and horticultural vermiculite.

3 Nip off side shoots close to the base and dip the ends of the cuttings in hormone rooting powder. Stick them 1cm (½in) deep into the compost. Each pot should hold six cuttings.

4 Pack the pots into a seed tray and stand them in a cold frame in a sunny corner. Water thoroughly and replace the framelight.

5 Check the frame about once a week. Water again if the compost starts to dry out. Extra insulation will be needed only during unusually cold weather.

6 From mid-spring onwards, pull gently at a few cuttings from time to time to see if they have formed roots.

Conifers make effective screens, and their bold shapes and distinctive colours and textures contrast with broad-leaved plants and shrubs. Many of the popular garden forms can be grown from cuttings taken in early winter (recipe 38). Others, particularly the species grown by foresters or as specimen trees in arboreta, are grown from seed (recipe 23). They are slow starters, taking several years to establish the foundation from which later growth and development takes off

• • • • • • • • •

117

7 As soon as roots are detected, drench thoroughly with a high-potash liquid feed (5×) and raise the front of the frame to provide ventilation during the day. Close it at nightfall.

8 When the rooted cuttings start to produce new growth, pot them individually into 7cm (2½in) pots and return them to the cold frame, with the front kept continuously raised.

9 In mid-summer take the pots out of the cold frame and stand them out in a sheltered corner.

10 During mid-autumn line the plants out 20cm (8in) apart in a nursery bed, and mulch with a 2cm (¾in) layer of composted bark.

11 Grow them on for another two or three years, before planting them in the garden.

POSSIBLE PROBLEMS

SYMPTOMS	CAUSE	REMEDY
Cuttings fail to form roots.	a) Conditions too cold wet in the winter. b) Ineffective hormon rooting powder.	Take care not to overwater. Add extra protection during periods of severe frost. Replace supply annually. Store in dark; if possible in a refrigerator.
Rooted cuttings grow poorly after potting up.	Lack of nutrients in compost.	Feed at fortnightly intervals with a high-potash liquid feed (2×).

GROW-YOUR-OWN DWARF CONIFERS

Many other conifers can be propagated in a similar way, including those conifers of restricted growth (CORGS) – often referred to as dwarfs or miniatures. Try cultivars of the following: Japanese cypress; Lawson's cypress and other species of *Chamaecyparis*; western red cedar and other species of *Thuya*; yew. (Junipers are not included in this list. Cuttings from them produce roots more successfully when taken in mid-summer.)

EXTREME ECONOMY: CUTTINGS FROM A SINGLE BUD

22

In 1973 a colleague in the USA sent me a bundle of prunings from a dozen different grape vines growing in upper New York State. I cut them into short lengths, each with a single bud, and laid them on grit in pots on a heated propagating bench. Most grew into young vines and I planted them along a pergola in my garden.

Two years later I looked forward to their first crop. It happened to be that never-to-be-forgotten, scarcely-to-be-repeated, summer of 1976 – and the vines revelled in the heat and drought. A seedless, green grape called 'Himrod' was ready by early August, and one after another they continued to produce deliciously flavoured, sweet grapes until the end of October. I thought I was on to something really worth while, but the following summer's cooler, moister, more normal weather showed me how wrong I was. It would be years before they matched their performance on their triumphal début.

As an attempt to introduce new varieties of grapes it was a failure. But the cuttings had grown and there are other plants, with a proven record of success in our gardens, that we can propagate from no more than a single bud. They include tempting prospects like the large-flowered clematis hybrids, camellias, lilies and roses.

Roses Grown from a Single Bud   RECIPE 39

```
TAKE CUTTINGS: Mid- to late summer
FEED: Early autumn and late winter
POT UP: Mid-spring
INTO NURSERY: Mid- to late summer (year two)
PLANT: Late autumn to early winter (year two)
PRUNE HARD: Early spring
FIRST FLOWERS: Mid- to late summer (year three)
```

OTHER SHRUBS THAT CAN BE GROWN FROM CUTTINGS CONSISTING OF A SINGLE BUD

Camellia
Clematis
Fatsia
Grape vine
Ivy
Mahonia

The easiest way to grow a rose is from hardwood cuttings taken in late autumn (see page 112). By then the flowers are probably faded memories, most of the leaves will have fallen and one bush can readily be confused with another. Single bud cuttings are taken in the summer so, when you like the look of a rose in a friend's garden, all you need ask for, to grow it yourself, is one short shoot. You can be sure it's the plant that you admired, even if its name has been lost and forgotten.

Roses and clematis can be propagated from very small cuttings.

Cuttings can be made from a single node, containing one bud (roses) or two (clematis). Although these small cuttings are economical – a number can be made from each shoot – they need more care, and take longer to grow into self-sufficient plants, than larger cuttings

—— WHAT TO DO ——

1 In late summer, cut off young shoots that have almost finished growing and begun to go woody – the smallest the thickness of a pencil, the largest about twice that size. They will have leaves, and buds in the angle where each leaf joins the main stem.

2 Fill 7cm (2½in) square pots with a 50/50 mixture of grit and perlite.

3 Slice the shoots into sections with a knife; cutting 2 or 3mm above each node to produce short, peg-like bits of stem with a leaf and bud at the top of each. Do not use the immature, upper third of the shoots.

4 Nip off the outer leaflets on each cutting, and dip its base into hormone rooting powder. Stick the cuttings

into the compost with a dibber so that the buds lie just below the surface. Try to arrange the leaves so that they lie within the perimeter of the pot, with five cuttings per pot.

5 Pack the pots into a shallow box or seed tray in a cold frame. Water thoroughly, damp down the inside of the frame and replace the framelight.

6 Keep the framelight closed, draping an extra sheet of bubble polythene over it on sunny days. Spray the foliage and surfaces within the frame to maintain a high humidity.

7 Six weeks later drench with a high-potash liquid feed (5×). Each week pull gently at a few cuttings to check whether they have produced roots. When they do so move the pots into an adjacent frame with the front raised

20cm (8in) for ventilation.

8 Feed with liquid feed (1×) at monthly intervals until the leaves start to drop.

9 Overwinter in the cold frame. When the rooted cuttings begin to grow the following spring, pot them individually into 7cm (2½in) pots and return them to the ventilated cold frame.

10 Once the roots of the cuttings have formed a network encircling the compost in the pots, plant them out 30cm (12in) apart in a nursery bed.

11 Mulch with 2cm (¾in) of composted or chipped bark. Keep them well watered. Some will be ready to plant in the garden that winter; others will need a second summer in the nursery bed.

Camellias were once believed to be tender; then they were thought to be faddy. Some certainly seem to flower rather shyly. Anyone who wonders whether camellias are for their garden should try 'Donation' first. If it fails forget about the others. This beautiful shrub can be grown from single bud cuttings (recipe 39), and develops steadily into a vigorous upright bush, with dark green glossy foliage. Best of all, it starts to flower while still small, and can be relied on to produce a grand display every year once it is established

. . . .

121

P O S S I B L E P R O B L E M S		
SYMPTOMS	**CAUSE**	**REMEDY**
Cuttings fail to form roots.	a) Ineffective hormone rooting powder. b) Cuttings too immature.	a) Replace rooting powder annually. Store in a dark, cool place; preferably refrigerated. b) Take cuttings only from firm green shoots, that have grown earlier in the year.
Cuttings fail to form shoots before the winter.	Summer dormancy.	Encourage shoots to grow by warmth, feeding and humidity. Nevertheless, many will not start to grow till the following spring.
Rooted cuttings die during the winter.	Stressed by wet and cold, made worse if cuttings are starved.	Keep leaves on cuttings green and active for as long as possible*. Do not overwater. Add extra insulation to frame in cold spells.

*These cuttings will grow faster, and be more likely to survive severe weather if space can be found for them on a heated bench in a greenhouse during the winter.

RECIPE 40 — Bergenia Slices for Maximum Production

TAKE CUTTINGS: Mid-summer
FEED: Mid-autumn and early spring
POT UP: Mid- to late spring (year two)
INTO NURSERY: Mid- to late summer (year two)
PLANT: Early to mid-spring (year three)
IN FLOWER: Early to mid-spring (year four)

Megaseas, elephant's ears, bergenias, call them what you will, are plants which people either love or find deadly dull. Not many years ago they drew attention to themselves only when tattered leaves of *Bergenia cordifolia* sprawling in a neglected corner produced a few puce flowers in spring. Now, there are close to fifty kinds to play with and their good points are more appreciated. The snag is that, although they eventually spread to cover quite large areas, they can be slow in doing so. If you depend on division to increase a new variety, it will take several years to put together enough plants for a good group.

The thick, root-like stems (rhizomes) of bergenias are covered with the shaggy remnants of old leaf bases, and, among them are hidden buds sitting in the angles where the leaves once joined the stems. These buds live on, forgotten for years, but each can grow into a new plant, given the chance. A snoop round a garden centre's stocks in mid-summer is often a good starting point. You may be lucky enough to find a plant which has grown old and ugly in its container, with two or three straggling stems and a tuft of leaves on each. Or you can use a plant from your own garden or beg one from a friend – the longer and the more straggly it is the better.

Most, like those listed below, are propagated by slicing into sections the upright stems on which the flowers are borne. Cuts should be made immediately above each leaf (cf. roses), and the cuttings stuck in the compost so that the bases of the leaves are just buried.

Dames violet
Foxglove
Lobelia, especially
   L. cardinalis

WHAT TO DO

1 Find a well grown bergenia plant in mid-summer. Cut off the clusters of leaves at the ends of the shoots.
2 Fill 7cm (2½in) square pots with a 50/50 mixture of grit and perlite.
3 Cut the rhizomes into 1cm (½in) thick slices; bury them in the cutting compost in the pots with their upper surfaces just beneath the surface. Approximately five slices should go into each pot.
4 Pack the pots into a seed tray in a cold frame. Water thoroughly and replace the framelight.
5 Check the pots occasionally and water them if they start to dry out.
6 When young shoots appear through the compost, drench with a high-potash liquid feed (5×).
7 Keep the cuttings in their pots in the frame through the winter, ventilated in all but severe weather.

a)

b)

c)

Making the most of bergenias. Bergenias can be propagated a) by division; b) by using their shoots as cuttings; c) by slicing their rhizomes into sections, each containing one or two dormant buds

8 Feed (2×) again in early spring, and in late spring pot each of the young plants which have developed into a 7cm (2½in) pot. Put them back in the frame.

9 Grow them on till mid-summer with plenty of ventilation. Then take the pots out of the frame and stand them in a sheltered place out of doors.

10 A fortnight later plant them out in a nursery bed. Mulch well.

11 Some may be big enough to plant in the garden late in autumn, most will need a second year in the nursery.

POSSIBLE PROBLEMS

SYMPTOMS	CAUSE	REMEDY
Developing shoots grow slowly in their pots.	Temperatures too low.	Erect the frame in a sunlit corner and keep it closed while shoots are developing. If space can be found, bring the pots containing the young plants into a greenhouse and stand them on a heated bench through the winter.

RECIPE 41 **Lilies Grown from Scales**

SCALE: Early winter
FEED: Late winter
POT UP: Mid-summer
FEED: Early and mid-spring (year two)
INTO NURSERY: Mid-summer (year two)
PLANT: Mid- to late summer (year three)
FIRST FLOWERS: Mid- to late summer (year four)

My garden in the Carse of Gowrie had been a chicken farm, and their pecking and scratching had decimated most of the plants long before it became mine. Magnificent among the survivors was a clump of lilies, with whorls of deep green leaves on strong stems and clusters of thick-petalled orange Turk's caps. I had never seen *Lilium hansonii* grow better; year by year the bulbs grew stronger and more numerous. I tried some elsewhere, but they failed to do well. Then one day I unearthed a chicken bone on the edge of the clump. First one, then another and another until it became clear that the lilies had been planted above a heap of bones – the burial-pit of some chicken-raising disaster!

Perhaps that pit was the key to growing lilies successfully. I've never seen them do so well since but I'm ever hopeful, and my greenhouse and frames now contain hundreds of small bulbs in preparation for yet another attempt to colonise a garden with these beautiful plants. A few are seedlings, but the great majority have been grown from the scales which make up a lily bulb. This is such an easy way to propagate them that I can never resist the temptation to remove a few scales every time I buy a bulb.

WHAT TO DO

1 During the winter, take a lily bulb and remove decaying or damaged outer scales. [The lily scales in this recipe are removed in wintertime; bulbs on sale then are a good starting point from which to produce more. Scales can be removed at any time of the year. In a garden, bulbs can be dug up when they have finished flowering, then scaled and replanted. In the summer there is no need to use a greenhouse; they can go straight into a frame.]

2 Gently break off the scales one by one, starting on the outside. Twist each slightly to the side and downwards, snapping it off at the point where it joins the central axis of the bulb. At least a third of the scales can be removed without wrecking the bulb.

3 Fill a seed tray with a 50/50 mixture of grit and perlite.

4 Stick the scales, base down, into the compost so that about two-thirds of each is below the surface. A tray holds about sixty scales – six rows of ten.

5 Stand the seed tray on a heated bench in a greenhouse and water thoroughly.

6 When the first green leaves appear, drench with a high-potash liquid feed (3×).

OTHER BULBS WHICH CAN BE PROPAGATED FROM SCALES

Fritillary
Nomocharis

Lily bulbs can be grown from scales.

A lily bulb is made up of fleshy scales lightly attached in a spiral around a very compressed stem. When these scales are snapped off, and half buried in gritty compost, small bulbils develop along the broken edges at the base of each scale. These develop into bulbs able to produce flowers after two to four years

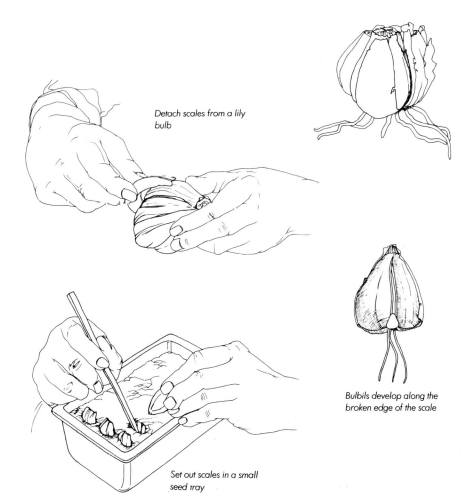

Detach scales from a lily bulb

Set out scales in a small seed tray

Bulbils develop along the broken edge of the scale

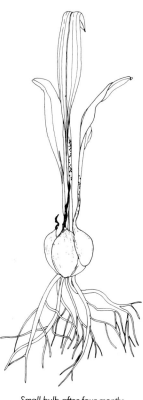

Small bulb after four months

125

Two-year-old plants of Lilium
martagon *and* L. pyrenaicum

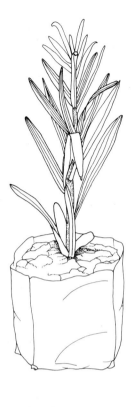

7 In mid-spring move the tray to a
ventilated cold frame; continue to
feed at monthly intervals.

8 In late summer pot the bulbils
individually into 7cm (2½in) square
pots and put them back in the cold
frame. It does not matter if most of
the leaves fall off while you do this.

9 Keep the pots in the cold frame
through the winter; early in spring,
drench with a feed (1×).

10 Grow the small bulbs on till mid-
summer, feeding them once a month;
then knock them out of their pots
and line them out in a nursery bed,
10cm (4in) apart.

11 One year later, as the leaves die
down in late summer, dig them up
and move them to permanent
positions in the garden.

POSSIBLE PROBLEMS		
SYMPTOMS	CAUSE	REMEDY
Bulbils fail to form at base of scales.	Scales broken off too high up.	Bulbils are formed from tissues at the very bottom of each scale. If the scales are broken in half, or part way up, they may not produce any bulbils.
Bulbils fail to come up after being put into pots.	Lilies renew root growth during late summer. During the winter small bulbs may produce no roots and fail to establish.	Pot the bulbils up, and transplant them, during late summer when their roots are growing actively. Avoid disturbing them during the winter.
Young bulbs die during the winter.	Killed by frost and wet.	Immature bulbs in small pots are much more vulnerable than large bulbs in the garden. Protect them in a frame or move them under the benches of a greenhouse for the winter.

MAKING CUTTINGS FROM ROOTS

23

Tame rabbits enjoy dandelion leaves, and as a boy I would go round the garden pulling the tops off the dandelions, returning to the hutches with handfuls of greenery. I soon discovered that the plants seemed to thrive on this treatment – in spite of having half their roots torn off with their tops – and went back time and again for more.

The dandelion's enthusiastic production of new shoots from any roots left in the ground is a frustratingly familiar part of gardening. Yet the idea that roots (of many plants) might be used as cuttings, strikes people as unlikely – even a little bizarre. It remains one of the lesser used methods of propagation; something to fall back on when seeds or division, or cuttings made from the shoots, are either too slow or too unreliable.

Cuttings from the Roots of Drumstick Primulas   RECIPE 42

Primula denticulata produces seeds which germinate freely and quickly grow into small plants. Why look for any other way to propagate it? The answer lies in the pleasure we take in flowers that are bigger, better or brighter than the usual. The usual is an attractive plant with round heads of lilac flowers. Our pleasure lies in heads of flowers that are not lilac and seem better because they are bigger. They may be white or crimson: dazzlingly white perhaps, or the deepest most covetable shade of crimson, in large, dense and immaculately symmetrical spheres. They too produce seed, and raising seedlings from them is no problem. But few, if any, will turn out to be as fine as their parents.

Another way to propagate our prize specimens could be by division: separating one crown from another to produce identical copies. But drumstick primulas do not have generous natures and this will be a slow process. If we want more than a few plants, fairly quickly, the answer is to take cuttings we need, but cannot obtain from the tops, from the roots.

———— WHAT TO DO ————

1 Dig up plants to be propagated in late winter, and wash off clay and clinging soil.

2 Cut off a few of the long thick white roots close to their origin from the crown. Up to half can be taken without seriously reducing the prospects of

the parent plant, which should be replanted at once.

3 Slice the roots into sections 3cm (1¼in) long. Lay them out in a row with their lower ends pointing towards you.

4 Fill 7cm (2½in) square pots with a

127

Roots of the drumstick primula can be used to make cuttings.

Large roots are cut into 3cm (1¼in) long sections which are inserted vertically into a gritty compost, and then just covered with a topping of the compost. Small plants develop from the submerged tips

50/50 mixture of grit and perlite.

5 Make holes in the compost with a dibber, and drop each section of root in with its top just below the surface of the compost. Nine sections to a pot.

6 Pack the pots into seed trays, or shallow boxes, and put them into a cold frame. (When space is available stand them on a heated bench in a greenhouse. They will do very well on it, and develop faster than in a cold frame.) Water thoroughly and replace the frame light.

7 Open the framelight on sunny days, but keep closed at other times. Add extra insulation on very cold nights.

8 When the first green shoots appear, drench with a high-potash liquid feed (3×).

9 Feed at fortnightly intervals (1×) till the developing plants have two or three fully formed leaves. Then pot each individually into a 7cm (2½in) pot. Put the plants back in the cold frame and grow them on with plenty of ventilation.

10 By mid-summer the plants can be stood out in the open. Plant them in the garden as soon as the young plants grow large enough to hold their own.

OTHER PERENNIAL PLANTS WHICH CAN BE PROPAGATED FROM CUTTINGS MADE FROM THEIR ROOTS	
Bear's breeches	Mullein
Border phlox*	Oriental poppy
Eryngium	Pasque flower
Japanese anemone	Romneya

*This is a useful way to save phloxes which have become infested with eelworm. But, cultivars with variegated leaves, such as 'Norah Leigh' and 'Harlequin', come up with uniformly green leaves when propagated from root cuttings. Variegated-leaved plants cannot be reproduced in this way.

POSSIBLE PROBLEMS		
SYMPTOMS	**CAUSE**	**REMEDY**
Cuttings rot without forming shoots.	Too wet, too cold, or both.	Be careful not to overwater: very little will be needed until shoots start to to grow. Protect the cuttings with extra insulation on the frames in severe weather. Or lay a piece of bubble polythene directly over the pots till leaves appear.
Developing shoots wilt and then disintegrate.	Infected by water-borne fungi cf. damping off.	Water with tap water or clean out and disinfect storage tanks, water butts etc every six months.

This cluster of drumstick primulas have grown themselves from seed – and seed collected from any of these plants would also produce mixtures of white and lilac seedlings. Individual colours could be obtained by dividing the crowns – a slow process in this case – or by making cuttings from the strong white roots (recipe 42) found beneath the plants during the winter

24 CUT-COST REFILLS FOR POTS AND BASKETS

 I have been taught a new gardening term during the last year or two – 'Patio Plants'. I've no idea who coined the phrase or precisely which plants it covers. Most are tender perennials: marguerites, fuchsias, geraniums, verbenas etc. These are bright things, intended to make a splash in small beds, pots and other containers, that can be relied on to continue colourfully for months.

Practically none are hardy. They depend on almost frost-free accommodation to survive cold winters, and if we want to enjoy them every year we must spend time and money keeping them warm and watered, or buy again each spring.

That can be an expensive process – a hundred plants and more may be needed to freshen up even a modest patio – every year! But it is possible to buy a fraction of what you need and still make a display that looks as though you've spent a fortune. Some can be grown from seed. Buy a few plants of others in early spring and take cuttings from them at once to grow on. These tender perennial, so-called patio plants grow so quickly as the weather warms up that the handicap of their late start is soon made good.

RECIPE 43	**Busy Lizzies from Seed**

An earlier recipe for growing petunias (see page 59) stressed the advantages of waiting till late spring before sowing the seeds of bedding plants from tropical parts of the world – and none are more tropical than busy lizzies from the clove-scented island of Zanzibar. When the bedding is intended for the garden, late sowing works well because other plants flowering and growing in the borders keep us happy until it comes into flower. But nobody wants to wait long for plants in containers and baskets to give of their best – we all expect to see at least a bit of colour on them from the start.

OTHER PLANTS THAT CAN BE GROWN IN A SIMILAR WAY

African marigold
Ageratum
Basil
Begonia
Black-eyed Susan
  (*Thunbergia*)
Cock's comb
Coleus
Gazania
Lobelia (Prick
  these out in
  clusters – not
  individually.)
Morning Glory
Petunia
Salvia
Verbena

That means we must sow seeds that grow well only at high temperatures, before the winter is over, and when short, dark, chilly days and freezing nights are still with us. This is the moment when the womb-like, private warmth of a heated propagator becomes an asset. But remember that propagators are only temporary shelters, and are assets only if there is somewhere else light and warm to put the seedlings when they outgrow them. A space in a warm conservatory would do, or a greenhouse bench with soil-heating cables.

——— WHAT TO DO ———

1 Sow seeds in late winter on the surface of vermiculite in a 7cm (2½in) square pot. Plough them in.

2 Pack the pots into the propagator tray in a well-lit position, ie in a greenhouse or in a conservatory, and water them thoroughly.

3 Put the propagator lid on; open the ventilators and hang a thermometer through one so that its tip rests on the surface of the vermiculite. Adjust the thermostat control until the thermometer records a minimum temperature of 20°C (68°F).

4 Drape a sheet of bubble polythene over the propagator lid to maintain a high temperature during spells of cold weather.

5 As soon as signs of germination become visible, remove the lid of the propagator each day, provided air temperatures around the seedlings do not fall below 12°C (54°F).

6 Prick the seedlings out when they are large enough to handle. [Seedlings ready to be pricked out can be bought from many garden centres in small pots. This can be an economical way to avoid the early stages of production. *But* these seedlings have a very short 'shelf life' – make sure that they are still small when you buy them, and prick them out at once.] Unless a large number of one kind is needed, put them in small 7 or 9cm (2½ to 3½in) pots, with six or nine seedlings in each.

7 Put the pots in a warm conservatory or stand them on a heated bench, where temperatures of 15°C (59°F) or more can be maintained day and night. A canopy of bubble polythene can be used to protect the seedlings at night and on cold days.

8 When the leaves of neighbouring seedlings overlap, pot them individually into 7cm (2½in) square pots, returning them to the greenhouse or conservatory.

9 In late spring the plants can be moved out to a cold frame. Keep this closed or barely ventilated for the first week unless the weather is exceptionally mild. Use extra insulation when night temperatures fall below 5°C (41°F).

10 Plant in their permanent positions in baskets, troughs or other containers in early summer when all prospects of frost are past.

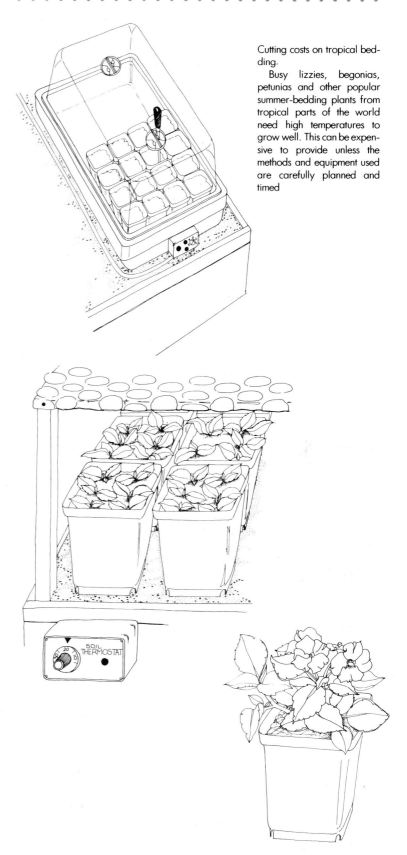

Cutting costs on tropical bedding.

Busy lizzies, begonias, petunias and other popular summer-bedding plants from tropical parts of the world need high temperatures to grow well. This can be expensive to provide unless the methods and equipment used are carefully planned and timed

POSSIBLE PROBLEMS		
SYMPTOMS	**CAUSE**	**REMEDY**
Seedlings fail to germinate.	Poor quality seed or conditions too cold.	These seeds can lose viability rapidly. Always use freshly bought seed. Temperature should be at least 20°C (68°F). If it drops below 15°C (59°F), germination may be reduced.
Newly emerged seedlings elongate and go pale yellow-green.	Etiolation, due to insufficient light combined with high temperatures.	Keep the propagator in as light a position as possible, especially when the seedlings start to germinate. The first couple of days are critical.
Seedlings fall over and rot away.	Damping off, due to overwatering or low temperatures.	Water with tap water, and only when really necessary. Maintain minimum temperature of 15°C (59°F) through-out the seedling period.
Young plants in frames fail to grow; foliage shrivels or becomes deformed.	Temperature too low.	Avoid early transfers to frames. Close the ventilators at night and during cold days. Use extra covers on cold nights.

RECIPE 44   *Viola* 'Prince Henry' for Winter Colour

Flower power is everywhere in the summer; winter is the testing time, especially in containers and small borders close to the house. Bulbs are a standby but few perform till spring is sprung, and there can be long periods of earthy drabness before then. Very few plants come out fresh and attractive in the autumn, and keep going through all but the coldest winters.

Winter-flowering pansies, particularly some recently introduced strains, manage it in most winters. Even better are two close relatives: the purple and crimson viola 'Prince Henry', and the bright orange-yellow 'Prince John'. Both, grown from seed, can be used to fill the holes and empty spaces as other plants give up the ghost in autumn.

(Opposite) Pots and containers provide space for a special kind of gardening: one where a bright display for the longest possible time is a high priority, and also one that tempts new ideas and fresh approaches every time the plants need to be renewed. But renewal and change can be expensive when every plant has to be paid for. Fortunately, the great majority of the bright young things that make container planting beautiful can be home-grown either from seed (recipe 43, 44 or 45) or from cuttings taken from a small stock of plants (recipe 46), bought to provide a nucleus for rapid expansion

——— WHAT TO DO ———

1 Sow seeds in late summer on vermiculite in a 7cm (2½in) pot; plough them lightly beneath the surface.

2 Pack the pots into a seed tray in a cold frame. Water thoroughly and close the lights. They could share space with cuttings in a propagating frame.

3 When the seedlings are large enough,

prick them out. A standard tray will hold sixty plants; or put them into 9cm (3½in) pots, nine to a pot. Return them to the cold frame, water thoroughly and after four days raise the front 20cm (8in).

4 Two weeks later remove the frame light or put the plants in a sheltered place out of doors.

132

5 Pot the young plants into 7cm (2½in) pots when their leaves touch, and stand them outside.

6 Grow them on. If their leaves reach the edges of the pots before they are planted, feed with a high-potash liquid feed (2×) every fortnight.

7 Plant in containers, baskets, borders etc, whenever space becomes available.

OTHER PLANTS WHICH CAN BE GROWN IN A SIMILAR WAY

Winter-flowering pansies in a variety of colours, and *Viola* 'Prince John'.

POSSIBLE PROBLEMS

SYMPTOMS	CAUSE	REMEDY
Young plants do not begin flowering by the time they are planted.	Sown too late: short of nutrients after potting up.	Optimum sowing date will vary in different places: next time sow a fortnight earlier. Take care to make sure that plants do not go short of nutrients. Start feeding potted plants before they show symptoms of need.

Growing Geraniums from Seed
RECIPE 45

I used to follow the advice on the backs of packets of geranium (*Pelargonium*) seeds and sow them in mid-winter. Then I came across an article in the trade press that told me that some growers found them more economical and easier to grow when sown in early autumn. It changed my life!

There were no sickly, newly emerged seedlings to be nursed through the darkest days of winter; no longer did the first flowers appear so late that threats of frost crimped their display when they were beginning to look their best. Instead, the young plants grew sturdy by mid-winter and resilient to cold and poor light. They started to flower in early summer, and by autumn had been colourful for months.

It sounds like a long and expensive way to grow geraniums. Longer it is, and careful planning is needed to make it worthwhile. Expensive it could be if a few geranium plants spent the winter in sole occupation of a heated greenhouse. The secret is to make sure the greenhouse is filled with a variety of plants and cuttings, using soil-heating cables in the benches, backed up by a covering of bubble polythene during cold spells, as the only source of artificial heating.

Carnation
Gazania
Pelargonium
 (scented-leaved)

Autumn-sown geraniums will bring a smile to your face in mid-winter.

Geranium plants grown from seeds sown in the autumn are large enough by mid-winter to survive the effects of poor light and low temperatures. Seedlings newly emerged from winter-sown seeds are more likely to die or be damaged by these conditions

WHAT TO DO

1 Sow seeds in early autumn on vermiculite in 7cm (2½in) pots. Plough the surface lightly so that the seeds sink just below the surface.

2 Put the pots on a bench in the greenhouse above soil-heating cables, water thoroughly and cover with an expanded polystyrene tile.

3 Set the thermostat on the heating cables so that temperatures at the surface of the vermiculite do not drop below 20°C (68°F). A heated propagator could be used to germinate the seeds.

4 When the seeds germinate, remove the polystyrene tiles and switch off the heating cables.

5 When the seed leaves are fully expanded and the first true leaf is just visible, pot the seedlings individually into 7cm (2½in) pots.

6 Keep the plants in these pots during the autumn and winter on a bench above heating cables. Set the thermostat to maintain temperatures of about 10°C (50°F) at the bench surface. Cover with a sheet of bubble polythene on very cold nights.

7 Water sparingly all winter. The plants will grow slowly but steadily and develop into compact plants with a dozen or more leaves by early spring.

8 In early spring pot into one litre pots. They can be planted in hanging baskets at this stage if that is required. Water more frequently to encourage them to grow actively.

9 In late spring move the plants into a cold frame. Keep the fronts open except during periods of cold winds and on frost-threatening nights. Use extra insulation if necessary to exclude frost.

10 Plant out in containers or beds at any time from early summer onwards.

POSSIBLE	PROBLEMS	
SYMPTOMS	CAUSE	REMEDY
Seedlings emerge with malformed seed leaves.	Defective seed.	This is a common problem. Pot the better ones up with the others. Some will develop into normal plants.
Plants die during the winter.	Killed by cold and damp.	Plants will be killed by temperatures below –3°C (27°F). Use bubble polythene to protect from frost during cold spells. Do not over-water, and avoid watering at all during chilly, overcast weather.
Leaves go mouldy in the winter.	Damp and lack of ventilation.	Keep ventilators open day and night, except during frosty weather. Pick off dead and mouldy leaves by hand regularly.

Cuttings of *Verbena* 'Sissinghurst' in Spring   *RECIPE 46*

This is a recipe for those who want to avoid the trouble of growing seedlings and looking after plants through the winter. Instead, let the growers do the work for you; then buy a few stock plants in the spring and use them as a source of all the plants you will need for an abundant, cheerful display later in the summer.

——————— WHAT TO DO ———————

1 In early spring buy a few, bushy, well-grown plants. The more shoots the better.

2 Take all the cuttings you can – each need consist of only two joints. Keep the parent plants, as they will soon produce new shoots for more cuttings.

3 Fill 7cm (2½in) pots with a 50/50 mixture of grit and perlite.

4 Dip the ends of the cuttings in hormone rooting powder; stick them 1cm (½in) into the cutting compost.

5 Set the pots out on a greenhouse bench above soil-heating cables (a heated propagator could be used) and water them thoroughly. Set the thermostat to maintain a temperature of 15°C (59°F). On cold nights protect the cuttings by laying a sheet of bubble polythene over them.

6 Roots will appear within ten days. Drench with high-potash liquid feed (3×), and seven days later pot up the rooted cuttings individually into 7cm (2½in) square pots. Return them to the heated bench.

7 Grow the plants on till late spring, with the ventilators open except in very cold weather; then move the plants to a cold frame or plant them in hanging baskets.

8 When the threat of frost has passed, the plants can be put outside and planted where they are to spend the summer.

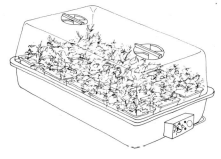

Patio plants can be expensive, but a few plants bought early in the season can be used as stock plants from which to produce a brilliant display later on.

Verbena 'Sissinghurst' is very easy to grow from small cuttings, which quickly grow into plants covered with flowers

'PATIO' PLANTS THAT CAN BE GROWN FROM CUTTINGS

Few develop so rapidly as the verbenas, but all will grow into plants well able to make a worthwhile contribution.

Anthemis cupaniana
Cineraria maritima
Felicia
Fuchsia
Gazania
Geranium
Helichrysum
Heliotrope
Marguerite
Osteospermum
Penstemon
Pelargonium
   (scented-leaved)
Phormium

P O S S I B L E P R O B L E M S		
SYMPTOMS	**CAUSE**	**REMEDY**
Plants fail to grow fast enough to make a display.	Conditions too cold.	The cuttings must make rapid progress during the first month, and need artificial warmth to do so. Unless this is adequate it is not worth attempting this recipe.

25 CAREFREE COLOUR FOR CONSERVATORIES

Conservatory gardening — under shelter, in the warm, close to the comforts of the house — has great appeal, especially in winter. To make the most of it, try growing plants to fill spaces, to replace tired oldies and to set up changing displays on the benches and at floor level. This bench has a mixture of temporary and semi-permanent plants contributing to the display including streptocarpus (recipe 49), temple bells, two ferns — an asplenium and a stag's horn — a tall abutilon, and a seedling palm tree. Other plants, such as schizanthus (recipe 47) or autumn chrysanthemums (recipe 48) could replace some of these at different times of the year

One of the great pleasures of Kew when I first arrived there, was Greenhouse No 4. Those who entered it stepped into a world of such colour and fragrance, that few resisted an intake of breath and an exclamation of delight. This was not the orchid house, or the place where fantastic cacti grew; it was not one of the historic greenhouses – the Palm House and the Temperate House were both close to collapse. It was a house maintained as an 'old-fashioned' Victorian – perhaps Edwardian – conservatory.

This inspiration was axed a few years later in a bid to divert resources to more scientific aspects of Kew's work. The imagination and the hard work which had maintained the high drama of the display were absorbed elsewhere and a duller but less demanding collection of greenery took its place. It is a common story. How many of our own conservatories match the dreams which led to their construction? Some do so for a year or two; then begin to fill with worthy but easily grown plants that lack the sparkle, colour and fragrance that can produce a magical other world in the depths of winter.

Sometimes this happens because the owners lose interest, but it is also an inevitable result of the nature of conservatories. Restricted within the confines of the structure, it is impossible to make the most of them; to go on changing the display, and enjoying fresh sensations as different flowers come and go. Space must also be found for plants that are past their best and need to rest; for future stars, and for the cuttings and seedlings from which they will arise. Anyone whose hopes for their conservatory extend further than a passive background of greenery will need a small greenhouse, a frame or two and a little space for a nursery besides.

136

Schizanthus from Seeds   *RECIPE 47*

SOW: Early autumn
POT UP: Mid- to Late-autumn
FEED: Throughout winter
RE-POT: Early spring
IN FLOWER: Mid- to late spring

Winter and early spring can be the best of times in a conservatory, and the times when its flowers are most enjoyed. Not surprisingly, many of the most valued occupants are plants that grow and flower naturally during the winter. The butterfly plant is one of these and shares with others an inclination to thrive at low temperatures, so long as they do not drop below freezing point, and to suffer less than gladly when kept too warm.

———— WHAT TO DO ————

1 Sow seeds in early autumn, sprinkling them over vermiculite in a 7cm (2½in) square pot, and ploughing them under the surface with a blunt dibber.

2 Put the pot on an unheated bench in a greenhouse and cover with a flat piece of expanded polystyrene.

3 When the seedlings grow large enough to handle, pot them individually into 7cm (2½in) pots.

4 Stand them on a heated bench in a greenhouse with the thermostat set to maintain a temperature of 10°C (50°F) at its surface. Cover the plants with a sheet of bubble polythene on cold nights. Keep the ventilators open on all but the coldest nights.

5 The plants should grow slowly and steadily throughout the winter. Feed once a month with a high-potash liquid feed (1×).

6 In late winter, transfer the young plants to one litre pots, and support them with hazel or similar twiggy shoots round the edge of the pot.

7 Continue to grow them under cool conditions with plenty of ventilation.

8 They should grow into bushy, upright plants, covered with flowers from mid-spring onwards.

Grow schizanthus from seed to brighten a conservatory in spring.

Schizanthus — sometimes called 'the poor man's orchid' — can be grown from seed sown in early autumn. It grows best with plenty of ventilation and needs little warmth, producing spectacular flowering plants during the spring

137

OTHER CONSERVATORY PLANTS THAT CAN BE GROWN FROM SEED

Achimenes
Begonia
Busy Lizzie
Calceolaria
Carnation
Cigar plant
Coleus
Cyclamen
Eucalyptus
Gerbera
Geranium
Heliotrope
Hypoestes
Kalanchoe
Ornamental pepper
Primula
Solanum

The seasons in which they should be sown and their treatment afterwards vary from one plant to another.

POSSIBLE PROBLEMS

SYMPTOMS	CAUSE	REMEDY
Seedlings die after potting into 7cm (2½in) pots.	The seedlings are vulnerable at this stage and need very careful handling.	The seedlings should be moved while still very small, to minimise disturbance and damage to their roots.
Young plants rot at ground level during the winter. Leaves start to go mouldy.	Overwatering; damage from cold or lack of ventilation leading to attack by grey mould.	Water sparingly and only during sunny periods. Ventilate very freely, but make sure that temperatures around the plants do not fall below freezing point.
Plants grow tall and straggly.	Too warm during the winter.	Try to avoid temperatures higher than 15 to 20°C (59 to 68°F) except in very sunny weather. During dull conditions temperatures of 5 to 10°C (41 to 50°F) are quite high enough.

RECIPE 48 — Chrysanthemums from Cuttings

TAKE CUTTINGS: Late winter
POT UP: Early spring to early summer
FEED: Mid- summer and early autumn
IN FLOWER: Mid-autumn to mid-winter

I always enjoyed growing chrysanthemums on the nursery. They followed the tomatoes, a filthy crop to work with and in such wretched condition by the end of their season we were glad to see them go. We carried the pots of chrysanthemums in from outdoors and they filled the houses with fresh, fragrant foliage, and a succession of richly coloured flowers throughout the autumn and first half of the winter.

Now they are a twelve-month-a-year crop. But amateurs and fanciers still like to grow them as one of the supreme promises of autumn.

WHAT TO DO

1 After the plants finish flowering, cut back all growth to 10cm (4in) above ground level. Tie labels to the stumps and stand the pots under the bench in a cold greenhouse, or in a frost-free frame. During this resting stage they should be kept at low temperatures, without risking damage from frost – minimum temperature 3°C (37°F).

2 In late winter, knock the plants out of

their pots and pack them into shallow trays, eg tomato boxes with a little potting compost; top dress the roots with a thin 1.5cm (½in) layer. Put the trays on a bench in the greenhouse. Water to keep compost moist but not wet.

3 Take cuttings from basal shoots 7 to 10cm (2½ to 4in) long. When available, use ones that grow from short underground shoots a little way from the old stems, as these may already have a few small roots.

4 Fill 7cm (2½in) square pots with a 50/50 mix of perlite and potting compost.

5 Nip off the leaves on the lower halves of the cuttings. Stick them 2cm (¾in) deep in the cutting compost with a dibber. Four cuttings in each pot.

6 Put the pots in a propagating frame on the greenhouse bench. Leave it open for three to four hours, before watering and closing the frame.

7 Keep the frame closed for the first two weeks. Temperatures around 10°C (50°F) are sufficient. Then open the front about 5cm (2in) to provide a little ventilation. Gradually increase ventilation. Spray very lightly; water seldom and sparingly.

8 When the cuttings are well rooted, pot them individually into 7cm (2½in) square pots and stand them on the bench.

9 Once the roots have encircled the compost in the pots, move them into a cold frame. A week later pot the young plants into one litre pots and pinch out the buds at the tips of the main shoots.

10 After a month stand them in the open, and about mid-summer pot up into five litre pots, sticking a 1m (3¼ft) bamboo cane into each pot.

11 Dig a trench and line out the pots so that half their depth is below soil

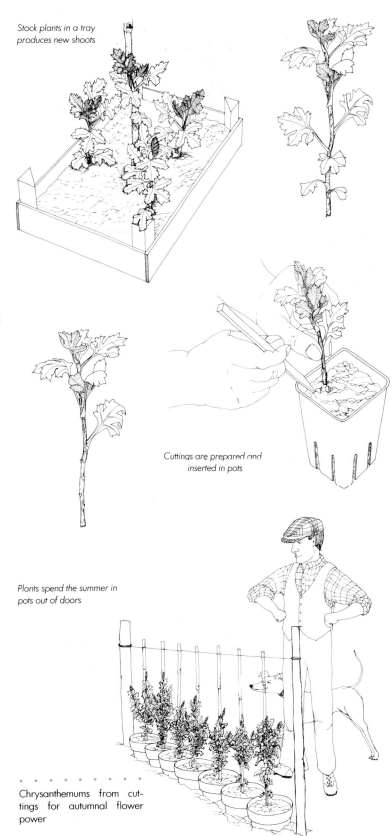

Stock plants in a tray produces new shoots

Cuttings are prepared and inserted in pots

Plants spend the summer in pots out of doors

Chrysanthemums from cuttings for autumnal flower power

139

OTHER CONSERVATORY
PLANTS THAT CAN BE
GROWN FROM
CUTTINGS

Abutilon
Achimenes
Camellia
Carnation
Christmas cactus
Cigar plant
Coleus
Datura
Fuchsia
Geranium
Heliotrope
Hoya
Hydrangea
Hypoestes
Kalanchoe
Lemon verbena
Marguerite
Myrtle
Passion flower
Pelargonium
Pittosporum
Plumbago
Setcreasea
Tibouchina
Tradescantia

The methods used vary
considerably from one plant
to another.

level. Refill the trench; drive a stake in at each end and tie the canes to a wire stretched between the stakes.

12 When the side shoots have grown 12 to 18cm (4½ to 7in) long, pinch out the tips of late varieties to encourage the plants to grow compact and bushy. Keep the pots well watered through the summer and, from early autumn, feed at fortnightly intervals with a high-potash liquid feed (1×).

13 Bring the plants into the conservatory when the flower buds start to show colour, or before the first frosts of winter.

14 Once under cover they should be grown as cool as possible, provided temperatures do not drop below freezing point, in a light, well-ventilated position, until the flowers are fully developed.

POSSIBLE PROBLEMS

SYMPTOMS	CAUSE	REMEDY
Base of the cuttings rot without forming roots.	Cutting compost too wet. Likely to be worse at high temperatures.	Allow cuttings to wilt before watering for the first time. Keep atmosphere moist and frame shaded. Avoid adding unnecessarily to water content of cutting compost.
Plants grow spindly, and produce few shoots during the summer.	Either the effects of drought or of starvation.	Never let the compost become so dry that the plants wilt. Begin feeding if lower leaves start to go pale, turn brown at the edges, or drop off prematurely.
Flowers go mouldy just before they start to open.	Atmosphere too damp, made worse by warmth and poor ventilation.	Avoid watering or damping down inside the conservatory on dull, windless, humid days. Keep ventilators open except during frosts. Night temperatures should not exceed 10°C (50°F).
Plants suffer from other fungal diseases or viruses.	Rust and mildew can attack the leaves. Viruses may result in misshapen flowers, reduced vigour, or mottling on leaves.	Pick off leaves affected by rust. Use proprietary sprays to control mildew (check that they are safe for chrysanthemums). Burn all plants infected with viruses and restock with virus-free plants.
Plants attacked by insects.	Aphids, leaf miner and earwigs cause damage to leaves and flowers, the former spread viruses.	Rub out aphids between finger and thumb*. Pick off leaves affected by leaf miner. Earwigs can be trapped in hay-filled flower pots on the tops of the canes.

*Use a systemic insecticide if attacks persist or increase in severity. Chrysanthemums are susceptible to numerous other pests and diseases, and specialist sources should be consulted for more information.

Streptocarpus from Leaf Cuttings

> TAKE CUTTINGS: Mid- to late spring
> POT UP: Mid-summer
> FEED: Late summer
> RE-POT: Early spring
> IN FLOWER: Late spring to early autumn

OTHER CONSERVATORY PLANTS THAT CAN BE GROWN FROM LEAF CUTTINGS

Begonia
Crassula
Echeveria
Saintpaulia
Sanseveria

Growing wild among rocks, by streamsides beneath heavy evergreen forest, it is not surprising that cape primroses are tolerant of dimly lit, dank conditions. This tolerance even extends to growing happily in the darker parts of a conservatory, where they thrive, tucked away behind and beneath more light-demanding plants.

They can be grown from seed; or plants which you already have can be propagated from cuttings made from their leaves.

———— WHAT TO DO ————

1 The leaves can be used to make cuttings whenever they are green and healthy.
2 Remove an entire leaf close to where it joins the plant and slice it across the midrib into sections 2 to 3cm (approx. 1in) wide. Line the sections out with their lower edges (the one closest to the plant) facing towards you.

3 Fill a half or quarter seed tray (depending on how many sections you have) with a 50/50 mixture of grit and perlite.
4 Stick the sections of leaf into the mixture so that they stand firmly on edge with their lower halves buried beneath the surface.

Cuttings made from the leaves of streptocarpus soon grown into new plants.

Fully grown, but not decaying, leaves are cut across the mid-rib to make 2.5cm (1in) deep sections. Small plants develop from cut surfaces near the mid-rib when the cuttings are put in a seed tray in a warm, shaded position

5 Stand the trays on a bench above soil-heating cables set to maintain temperatures between 15 and 20°C (59 and 68°F). Enclose them beneath a canopy of bubble polythene; and in sunny weather use a second sheet of bubble polythene to provide more shade.

6 Plantlets will develop close to the midribs, below the surface of the compost. When they appear, drench with a high-potash liquid feed (2×)

and repeat at fortnightly intervals. When their leaves grow to at least 2cm (¾in) long, move the plants individually to 7cm (2½in) pots.

7 Return them to the heated bench beneath the canopy and grow them on. Some of the cuttings may produce a few flowers in the autumn

8 Reduce watering during the winter and do not feed. Repot into one litre pots in mid-spring; they will flower during the summer.

POSSIBLE PROBLEMS

SYMPTOMS	CAUSE	REMEDY
Slow development of plantlets from leaf cuttings.	Too cool and too light.	Maintain temperature around 20°C (68°F), but do not let maxima exceed 25°C (77°F). Keep cuttings under shaded conditions at all times.
Plants lose their leaves prematurely.	High light intensities, more likely if plantlets are kept too dry.	Provide shading throughout the early period of development. Do not allow compost to dry out at this time.
Young plants rot during the winter.	Too cold; more likely to happen if the plants are kept too wet.	Maintain minimum winter temperature of 10°C (50°F), occasionally down to 5°C (41°F) for short periods. Water sparingly but do not allow compost to become dust dry.

RECIPE 50 **Freesias from Seeds**

CHIT/PRICK OUT: Late winter to early spring
POT UP: Early summer
FEED: Mid- summer to early autumn
IN FLOWER: Mid- autumn to late winter

Freesias can be grown from corms, planted in the autumn, or from seeds sown early in the spring.

——— WHAT TO DO ———

1 In early spring partially fill a half seed tray with perlite, water thoroughly and scatter the seeds evenly over the surface.

2 Put the tray on a heated bench in a greenhouse. Cover it with a polystyrene tile and set the thermostat to maintain temperatures of 15 to 20°C (59 to 68°F).

3 The seeds will germinate on the surface (chit) in ten to fourteen days. Select those with clearly visible well-formed roots and prick them out into 7cm (2½in) square plastic pots filled with standard potting compost. Each pot will hold six seeds spaced out in holes made with a dibber; fill in the holes with compost afterwards.

4 Put the pots on the greenhouse bench and water thoroughly.

5 A month to six weeks later, move the pots to a cold frame, and after two weeks transfer the contents of each into a one litre pot. Do not separate the young seedlings or disturb them more than necessary.

6 Leave the pots of seedlings in the frame throughout the summer, with the framelight raised to provide shade during the hottest part of each day. Support the seedlings with twiggy hazel sticks around the edge of the pots.

7 In mid-autumn the plants should be moved into a frost-free greenhouse, or straight into the conservatory. Grow on in cool conditions with as much ventilation as possible. The corms will produce a succession of flowers till the spring if the sprays are picked soon after they open and before they form seeds.

OTHER BULBOUS PLANTS THAT CAN BE GROWN IN SIMILAR WAYS

Ixia
Sparaxis
Tigridia

POSSIBLE PROBLEMS		
SYMPTOMS	CAUSE	REMEDY
Chitting seeds do not germinate, or roots appear deformed.	Most likely due to poor quality seed. Can be due to the effects of high temperatures.	Only use freshly bought, foil-wrapped seed. Make sure that temperature does not exceed 20°C (68°F) during the germination period.
Leaves wither or go brown on young plants during the summer.	Effects of high temperatures; will be accentuated if the plants suffer from drought.	Keep the plants as cool as possible during the summer, adding extra shade in heatwaves. Water freely in hot weather, and never allow the compost to dry out completely.

INDEX

Numbers in *italics* denote illustrations

THE
PROPAGATOR'S
HANDBOOK

Why spend money on plants when, with a few simple steps and with some expert advice, you can grow all you need at a fraction of the cost? *The Propagator's Handbook* explains simply and clearly how to divide plants, sow seeds and take cuttings using methods that really work. This book will transform your garden allowing you to grow an enormous variety of plants in the quantities of your choice.

The Propagator's Handbook is comprehensive in its coverage and includes propagation methods for annuals, perennials, shrubs, ferns, alpines, climbers, conservatory plants, bulbs and trees. In addition to providing the basic propagation principles for each plant group, the author also describes one simple method of propagation in a unique 'recipe' format. Sharing expertise gained from years of practical experience, Peter Thompson explains how to use propagation methods to 'save' tender plants which may be lost during winter and suggests ways of planning the propagation year to maximize interest in the garden.

Detailed illustrations further clarify the practical information and glorious photographs celebrate the finished results. Whether you want drifts of daffodils or acres of annuals this lively handbook is the key to successful propagating.

DAVID & CHARLES

£10.99